Lecture Notes in Mathematics

Edited by A. Dold, B. Eckmann and F. Takens

1406

L. Jacobsen (Ed.)

Analytic Theory of Continued Fractions III

Proceedings of a Seminar-Workshop,
held in Redstone, USA, June 26–July 5, 1988

Springer-Verlag

Berlin Heidelberg New York London Paris Tokyo Hong Kong

Editor

Lisa Jacobsen
Division of Mathematical Sciences
University of Trondheim
N-7034 Trondheim, Norway

Mathematics Subject Classification (1980): 01A55, 10A32, 10F35, 30B70, 33A30, 34A50, 40A15, 41A21

ISBN 3-540-51830-4 Springer-Verlag Berlin Heidelberg New York
ISBN 0-387-51830-4 Springer-Verlag New York Berlin Heidelberg

Printing and binding: Druckhaus Beltz, Hemsbach/Bergstr.
2146/3140-543210 – Printed on acid-free paper

PREFACE

This volume contains the proceedings of a research Seminar–Workshop held in Redstone, Colorado from June 26 to July 5, 1988. The topic of this workshop was Analytic Theory of Continued Fractions, and it was organized by William B. Jones, University of Colorado and Arne Magnus, Colorado State University.

This was the third workshop of its kind. The first one was held in Loen , Norway in 1981 (proceedings published in LN in M N^0 932), the second one in Pitlochry and Aviemore, Scotland in 1985 (proceedings published in LN in M N^0 1199). The idea is that workers in the field shall come together to exchange ideas and start cooperation. So, in addition to the presentation of our latest results, we have talks on ideas, half–finished projects, ideas that did not work, etc. Questions and comments, stupid or not, are encouraged.

This time analysis of papers from the turn of the century was one of the issues. Mathematicians like Helge von Koch, Jan Sleszynski and Julius Worpitzky had results and arguments which still deserve attention. Indirectly, these studies led to the study of separate convergence, a topic presented in a survey article in this volum, and to a historical article on Julius Worpitzky. In addition we continued discussions on nearness problems, the connection to Padé Approximants, and applications to number theory and differential equations. Interesting in this respect is for instance the use of Lange's δ–fractions to solve Riccati equations, and the question of finding thin subsets E_0 of an element region E whose best limit region V_0 is dense in the best limit region V for E. Also several other topics which are not reflected in these proceedings were discussed and will hopefully be published elsewhere.

It is evident that for workshops like this the location is important. Comfort, "isolation", a good seminar room and soothing nature are essential. We had all this in beautiful Redstone, and we are grateful to Nancy Lambert, manager of the Redstone Inn, for her hospitality. We also wish to express our gratitude to the Norwegian Research Council for Science and the Humanities, to the U.S. National Science Foundation, to the University of Colorado and to the universities of the respective participants for financial support. Finally we would like to thank Professor B. Eckmann for accepting this volume for publication.

Lisa Jacobsen

CONTENTS

LIST OF CONTRIBUTORS AND PARTICIPANTS

SANDRA CLEMENT COOPER
Department of Pure and Applied Mathematics, Washington State University, Pullman, Washington 99164–2930, USA

ROLF M. HOVSTAD
Nåkkves vei 5, L.1020, N–0670 Oslo 6, Norway

LISA JACOBSEN
Division of Mathematical Sciences, University of Trondheim, N–7034 Trondheim, Norway

WILLIAM B. JONES
Department of Mathematics, Campus Box 426, University of Colorado, Boulder, Colorado 80309, USA

L. J. LANGE
Department of Mathematics, University of Missouri, Columbia, Missouri 65201, USA

JOHN McCABE
Department of Applied Mathematics, University of St. Andrews, North Haugh, St.Andrews, Fife, Scotland, KY 169 SS, U.K.

OLAV NJÅSTAD
Division of Mathematical Sciences, University of Trondheim, N–7034 Trondheim, Norway

W. J. THRON
Department of Mathematics, Campus Box 426, University of Colorado, Boulder, Colorado 80309, USA

HAAKON WAADELAND
Department of Mathematics and Statistics, University of Trondheim, N–7055 Dragvoll, Norway

NANCY WYSHINSKI
Department of Mathematics, Campus Box 426, University of Colorado, Boulder, Colorado 80309, USA

ORGANIZERS

WILLIAM B. JONES, University of Colorado

ARNE MAGNUS, Department of Mathematics, Colorado State University, 121 Engineering, Fort Collins, Colorado 80523, USA

δ- Fraction Solutions to Riccati Equations

S. Clement Cooper

Department of Pure and Applied Mathematics

Washington State University

Pullman, WA 99164- 2930

1. Introduction

The objective of this paper is to introduce another method for solving scalar Riccati equations by continued fractions. The Riccati differential equation is of particular interest for several reasons: it is one of the simplest nonlinear ordinary differential equations, it is closely associated with a second- order linear differential equation, and it appears in many applications including general relativity [12,13], acoustics [8], systems theory [9], and invariant embedding [1].

Riccati equations have the convenient property that they are invariant (in a sense) under linear fractional transformations (lfts). More specifically, under an lft

(1.1)
$$y = \frac{\alpha(z)w + \beta(z)}{\gamma(z)w + \delta(z)}$$

a Riccati equation

(1.2)
$$y' = f_0(z) + f_1(z)y + f_2(z)y^2$$

is transformed into another Riccati equation

(1.3)
$$w' = \tilde{f}_0(z) + \tilde{f}_1(z)w + \tilde{f}_2(z)w^2 \ .$$

Since lfts play a fundamental role in the development of continued fractions [7], it is very natural to use continued fractions to solve Riccati equations. There has been a lot of recent interest in the use of continued fractions to solve Riccati equations as evidenced by [2,3,4,5,6,14,15].

Definition 1. *Let D be a (formal) differential operator. A continued fraction with n^{th} approximant $f_n(z)$ is said to be a formal solution of a differential equation $D[W(z)] = 0$ at $z = 0$ if*

(1.4)
$$\Lambda_0 \left(D[f_n(z)] \right) = O\left(z^{k_n} \right)$$

where $k_n \to \infty$ as $n \to \infty$. Here $\Lambda_0(f)$ denotes the Laurent series about $z = 0$ for a function f meromorphic in a neighborhood of zero.

In this paper a relatively new type of continued fraction, a δ- fraction, is used to find solutions to Riccati differential equations of the form

$$(1.5) \qquad R[W(z)] := A(z) + B(z)W(z) + C(z)W^2(z) - W'(z) = 0$$

and the form

$$(1.6) \qquad R[W(z)] := zA(z) + B(z)W(z) + C(z)W^2(z) - z^kW'(z) = 0$$

under the conditions that $W(0) = 0$, $A(z)$, $B(z)$, and $C(z)$ are analytic at $z = 0$, and in (1.6), $k \in \mathbf{Z}^+$.

The class of δ- fractions was introduced by Lange in 1981 [10]. This is a class of continued fractions whose members are finite or infinite continued fractions of the form

$$(1.7) \qquad b_0 - \delta_0 z + \frac{d_1 z}{1 - \delta_1 z} + \frac{d_2 z}{1 - \delta_2 z} + \cdots$$

where b_0 and d_n are complex constants, $d_n \neq 0$ for $n = 1, 2, \ldots$, and the δ_n are real constants equal to either 0 or 1. The δ- fraction is *regular* in case $d_{n+1} = 1$ whenever $\delta_n = 1$. Lange chose the name δ- fraction "because of the binary "impulse" nature of the sequence $\{\delta_n\}$ and the analogies, therefore, with the δ's in the Dirac delta function and the Kronecker delta symbol" [11]. His initial desire was to find a simple class of continued fractions of the form

$$(1.8) \qquad b_0(z) + \frac{a_1(z)}{b_1(z)} + \frac{a_2(z)}{b_2(z)} + \cdots$$

that had the following properties.

(i) The elements $a_n(z)$ and $b_n(z)$ are polynomials in z of degree ≤ 1.

(ii) The regular C- fractions

$$(1.9) \qquad d_0 + \frac{d_1 z}{1} + \frac{d_2 z}{1} + \cdots, \qquad d_n \in \mathbf{C}, \quad d_n \neq 0 \text{ if } n \geq 1$$

are in the class.

(iii) Given a power series

$$(1.10) \qquad L_0 = c_0 + c_1 z + c_2 z^2 + \cdots, \qquad c_n \in \mathbf{C}$$

there exists a unique member of the class that corresponds to L_0.

(iv) If L_0 represents a rational function in a neighborhood of $z = 0$, then its corresponding continued fraction terminates.

(v) For many classical functions analytic in a neighborhood of $z = 0$, the corresponding continued fraction has elements that can be expressed in closed form.

(vi) Convergence results can be obtained.

(vii) In many cases the approximants of the continued fraction corresponding to L_0 appear in the Padé table for L_0.

Many useful continued fractions, among which are C- fractions, general T- fractions, and P-fractions, satisfy some but not all of the requirements (i)- (vii). In [10,11], Lange has shown that the δ- fractions satisfy conditions (i) - (vi) and he has indicated that it is reasonable to expect that they also satisfy condition (vii). Two useful theorems from [11] that will be used in the sequel are the following.

Theorem 2. *Every regular δ- fraction (1.7) corresponds to a unique power series*

$$(1.11) \qquad L_0 = c_0 + c_1 z + c_2 z^2 + \cdots .$$

Conversely, for every formal power series (fps) (1.11) there exists a unique δ- fraction which corresponds to it. In the case of the infinite δ- fraction

$$(1.12) \qquad -\delta_0 z + \cfrac{d_1 z}{1 - \delta_1 z} + \cfrac{d_2 z}{1 - \delta_2 z} + \cfrac{}{} \cdots$$

the order of correspondence for the kth approximant is $k + 1$. In the case of the finite δ-fraction

$$(1.13) \qquad -\delta_0 z + \cfrac{d_1 z}{1 - \delta_1 z} + \cdots \cfrac{d_{n-1} z}{1 - \delta_{n-1} z} + \cfrac{d_n z}{1}$$

the order of correspondence for the kth approximant is $k+1$ if $0 \le k < n$ and ∞ if $k \ge n$.

Theorem 3. *A power series (1.11) is the Taylor series about the origin of a rational function if and only if there exists a finite regular δ- fraction that corresponds to it.*

In section 2, an algorithm is given for constructing δ- fraction solutions to Riccati equations of the forms given in (1.5) and (1.6). In section 3, theoretical results are given. The δ-fraction is shown to be a formal solution to the Riccati equation at $z = 0$ and the connection between the δ- fraction solution and the formal power series solution is given. The last theorem provides the link between the δ- fraction solution and a possible analytic solution. Section 4 is devoted to the computational aspects of the δ- fraction solution. In [3], C- fraction solutions were studied and some comments on the computational aspects were given. C-fraction solutions are compared to δ- fraction solutions and reasons are given for preferring δ- fractions over C- fractions.

2. Algorithm

In this section an algorithm is presented for generating regular δ- fraction solutions to initial value problems involving both nonsingular Riccati equations of the form

$$(2.1) \qquad R[W(z)] := A(z) + B(z)W(z) + C(z)W^2(z) - W'(z) = 0$$

and singular Riccati equations of the form

$$(2.2) \qquad R[W(z)] := zA(z) + B(z)W(z) + C(z)W^2(z) - z^kW'(z) = 0,$$

with $W(0) = 0$. The following definition identifies a class of Riccati equations in which every member is guaranteed a regular δ- fraction solution.

Definition 4. *A Riccati equation $R[W(z)] = 0$ is admissible if it satisfies the conditions in either (A.) or (B.).*

(A.) It is of the form (2.1) and $A(z)$, $B(z)$, and $C(z)$ are analytic at $z = 0$.

(B.) It is of the form (2.2) and it satisfies the following requirements.

$$(2.3) \qquad \begin{cases} (i)\, A(z),\ B(z),\ \text{and}\ C(z)\ \text{are analytic at}\ z = 0. \\[2mm] (ii)\, B(0)\ \text{and}\ C(0)\ \text{are not both zero.} \\[2mm] (iii)\, \text{The constant}\ k\ \text{is a positive integer.} \\[2mm] (iv)\, \text{If}\ k = 1, B(0)\ \text{is not a positive integer, and if}\ k > 1, B(0) \neq 0. \end{cases}$$

Theorem 5. *From every admissible Riccati equation it is possible to construct either a finite regular δ- fraction*

$$(2.4) \qquad -\delta_0 z + \cfrac{d_1 z}{1 - \delta_1 z} + \cdots + \cfrac{d_{n-1}z}{1 - \delta_{n-1}z} + \cfrac{d_n z}{1}$$

$\delta_k \in \{0,1\}$ for $k = 0,1,\ldots,n$ and $d_k \in \mathbf{C}\backslash\{0\}$ for $k = 1,2,\ldots,n$ or an infinite regular δ- fraction

$$(2.5) \qquad -\delta_0 z + \cfrac{d_1 z}{1 - \delta_1 z} + \cdots \cfrac{d_n z}{1 - \delta_n z} + \cdots$$

$\delta_k \in \{0, 1\}$ *for* $k = 0, 1, \ldots$ *and* $d_k \in \mathbf{C}\backslash\{0\}$ *for* $k = 1, 2, \ldots$.

Proof: First consider admissible Riccati equations of the form (2.1). A regular δ- fraction will be generated by a process involving the following substitutions,

$$(2.6) \qquad W_0(z) = -\delta_0 z + W_1(z), \qquad W_n(z) = \frac{d_n z}{1 - \delta_n z + W_{n+1}(z)}, \qquad n \in \mathbf{Z}^+.$$

A sequence of Riccati equations will also be generated from which the constants δ_k, $k \in \mathbf{Z}_0^+$ and d_k, $k \in \mathbf{Z}^+$, can be determined by forcing the equations to be admissible. Let $W(z) = W_0(z)$, $A(z) = A_0(z)$, $B(z) = B_0(z)$, and $C(z) = C_0(z)$. Starting with $W_0(z) = -\delta_0 z + W_1(z)$ define

$$(2.7) \qquad R_1[W_1(z)] := R_0[-\delta_0 z + W_1(z)] = R_0[W_0(z)]$$

so that

$$(2.8) \qquad R_1[W_1(z)] := A_1(z) + B_1(z)W_1(z) + C_1(z)W_1^2(z) - W_1'(z) = 0$$

with

$$(2.9) \qquad \begin{cases} A_1(z) = \delta_0 + A_0(z) - \delta_0 z B_0(z) + \delta_0^2 z^2 C_0(z) \\ B_1(z) = B_0(z) - 2\delta_0 z C_0(z) \\ C_1(z) = C_0(z). \end{cases}$$

Let $A_1^*(z) = A_0(z)$. Define δ_0 as follows:

$$(2.10) \qquad \delta_0 = \begin{cases} 0 & \text{if } A_1^*(0) \neq 0 \\ 1 & \text{if } A_1^*(0) = 0. \end{cases}$$

Clearly $W_0(z) \equiv 0$ is a solution in case $A_0(z) \equiv 0$, so terminate the process immediately in this case. Otherwise, continue.

The next transformation is $W_1(z) = \dfrac{d_1 z}{1 - \delta_1 z + W_2(z)}$ from which $R_2[W_2(z)]$ is defined

$$(2.11) \qquad R_2[W_2(z)] := \frac{(1 - \delta_1 z + W_2(z))^2}{-d_1} R_1 \left[\frac{d_1 z}{1 - \delta_1 z + W_2(z)} \right]$$

$$= \frac{(1 - \delta_1 z + W_2(z))^2}{-d_1} R_1[W_1(z)]$$

so that

$$(2.12) \qquad R_2[W_2(z)] = zA_2(z) + B_2(z)W_2(z) + C_2(z)W_2^2(z) - zW_2'(z) = 0$$

with

$$(2.13) \quad \begin{cases} zA_2(z) = 1 - \dfrac{1}{d_1}A_1(z) + 2\dfrac{\delta_1}{d_1}zA_1(z) - \dfrac{\delta_1^2}{d_1}z^2A_1(z) - zB_1(z) \\[2mm] \qquad\quad + \delta_1 z^2 B_1(z) - d_1 z^2 C_1(z) \\[2mm] B_2(z) = 1 - \dfrac{2}{d_1}A_1(z) + 2\dfrac{\delta_1}{d_1}zA_1(z) - zB_1(z) \\[2mm] C_2(z) = -\dfrac{1}{d_1}A_1(z). \end{cases}$$

Let $zA_2^*(z) = 1 - \frac{1}{d_1}A_1(z) - zB_1(z) - d_1z^2C_1(z)$. In order for $A_2(z)$ to be analytic at $z = 0$, d_1 must be determined by

$$(2.14) \qquad\qquad d_1 = A_1(0) \neq 0.$$

Note that if $\delta_0 = 1$, then $d_1 = 1$. Determine δ_1 as follows,

$$(2.15) \qquad\qquad \delta_1 = \begin{cases} 0 & \text{if } A_2^*(0) \neq 0 \text{ or } A_2^*(z) \equiv 0 \\[2mm] 1 & \text{if } A_2^*(0) = 0, \text{but } A_2^*(z) \not\equiv 0. \end{cases}$$

The process terminates if $A_2^*(z) \equiv 0$ yielding the finite δ- fraction

$$(2.16) \qquad\qquad -\delta_0 + \frac{d_1 z}{1}.$$

Otherwise, note that $A_2(0) \neq 0$, $B_2(0) = C_2(0) = -1$.

For $n \geq 2$, use the transformation $W_n(z) = \dfrac{d_n z}{1 - \delta_n z + W_{n+1}}$ to define $R_{n+1}[W_{n+1}]$ as

$$(2.17) \qquad R_{n+1}[W_{n+1}] := \frac{(1 - \delta_n z + W_{n+1}(z))^2}{-d_n z} R_n\left[\frac{d_n z}{1 - \delta_n z + W_{n+1}(z)} \right]$$

$$= \frac{(1 - \delta_n z + W_{n+1}(z))^2}{-d_n z} R_n[W_n(z)].$$

Thus,

$$(2.18) \quad R_{n+1}[W_{n+1}] := zA_{n+1}(z) + B_{n+1}(z)W_{n+1}(z) + C_{n+1}(z)W_{n+1}^2(z) - zW_{n+1}'(z) = 0$$

with

$$\begin{cases} zA_{n+1}(z) = 1 - \dfrac{1}{d_n}A_n(z) + 2\dfrac{\delta_n}{d_n}zA_n(z) - \dfrac{\delta_n^2}{d_n}z^2A_n(z) - B_n(z) \\[2mm] \quad + \delta_n zB_n(z) - d_n zC_n(z) \\[2mm] B_{n+1}(z) = 1 - \dfrac{2}{d_n}A_n(z) + 2\dfrac{\delta_n}{d_n}zA_n(z) - B_n(z) \\[2mm] C_{n+1}(z) = -\dfrac{1}{d_n}A_n(z). \end{cases}$$

(2.19)

Let

(2.20) $$zA_{n+1}^*(z) = 1 - \frac{1}{d_n}A_n(z) - B_n(z) - d_n zC_n(z).$$

In order for $A_{n+1}(z)$ to be analytic at $z = 0$, we must have

(2.21) $$d_n = \frac{A_n(0)}{1 - B_n(0)},$$

where it is noted below that $B_n(0) \neq 1$. Determine δ_n as follows,

(2.22) $$\delta_n = \begin{cases} 0 & \text{if } A_{n+1}^*(0) \neq 0 \ \text{ or } \ A_{n+1}^*(z) \equiv 0 \\[2mm] 1 & \text{if } A_{n+1}^*(0) = 0, \text{but } A_{n+1}^*(z) \neq 0. \end{cases}$$

The process terminates if $A_{n+1}^*(z) \equiv 0$ yielding the finite δ- fraction

(2.23) $$-\delta_0 z + \frac{d_1 z}{1 - \delta_1 z} + \cdots + \frac{d_{n-1}z}{1 - \delta_{n-1}z} + \frac{d_n z}{1}.$$

Otherwise the process continues. Note that $A_{n+1}(0) \neq 0$, $B_{n+1}(0) = -C_{n+1}(0) = B_n(0) - 1 = -n$. Notice also that $\delta_n = 1$ implies that $d_{n+1} = 1$ for $n = 1, 2, \ldots$, so the δ- fraction that is generated is a *regular* δ- fraction.

Now consider admissible Riccati equations of the form (2.2). The construction is very similar to that given in the first half, but here there are extra admissibility conditions that must be relied upon. The first substitution is $W_0(z) = -\delta_0 z + W_1(z)$ which is used to define

(2.24) $$R_1[W_1(z)] := R_0[-\delta_0 z + W_1(z)] = R_0[W_0(z)]$$

so that

(2.25) $$R_1[W_1(z)] := zA_1(z) + B_1(z)W_1(z) + C_1(z)W_1^2(z) - z^k W_1'(z) = 0$$

where

$$(2.26) \quad \begin{cases} zA_1(z) = \delta_0 z^k + zA_0(z) - \delta_0 z B_0(z) + \delta_0^2 z^2 C_0(z) \\[2mm] B_1(z) = B_0(z) - 2\delta_0 z C_0(z) \\[2mm] C_1(z) = C_0(z). \end{cases}$$

Let $zA_1^*(z) = zA_0(z)$. Define δ_0 as follows:

$$(2.27) \quad \delta_0 = \begin{cases} 0 & \text{if } A_1^*(0) \neq 0 \\[2mm] 1 & \text{if } A_1^*(0) = 0. \end{cases}$$

Clearly, $W_0(z) \equiv 0$ is a solution in case $A_0(z) \equiv 0$, so terminate the process immediately in this case. Otherwise, continue. Before proceeding, note that (2.3) guarantees that $A_1(0) \neq 0$.

For $n \geq 2$, use the transformation $W_n(z) = \dfrac{d_n z}{1 - \delta_n z + W_{n+1}(z)}$ to define $R_{n+1}[W_{n+1}]$ as

$$(2.28) \quad R_{n+1}[W_{n+1}] := \frac{(1 - \delta_n z + W_{n+1}(z))^2}{-d_n z} R_n \left[\frac{d_n z}{1 - \delta_n z + W_{n+1}(z)} \right]$$

$$= \frac{(1 - \delta_n z + W_{n+1}(z))^2}{-d_n z} R_n[W_n(z)].$$

A routine calculation produces

$$(2.29) \quad R_{n+1}[W_{n+1}] = zA_{n+1}(z) + B_{n+1}(z)W_{n+1}(z) + C_{n+1}(z)W_{n+1}^2(z) - z^k W_{n+1}'(z) = 0$$

where

$$(2.30) \quad \begin{cases} zA_{n+1}(z) = z^{k-1} - \dfrac{1}{d_n} A_n(z) + 2\dfrac{\delta_n}{d_n} z A_n(z) - \dfrac{\delta_n^2}{d_n} z^2 A_n(z) - B_n(z) \\[2mm] \qquad\quad + \delta_n z B_n(z) - d_n z C_n(z) \\[3mm] B_{n+1}(z) = z^{k-1} - \dfrac{2}{d_n} A_n(z) + 2\dfrac{\delta_n}{d_n} z A_n(z) - B_n(z) \\[3mm] C_{n+1}(z) = -\dfrac{1}{d_n} A_n(z). \end{cases}$$

Let

$$(2.31) \quad zA_{n+1}^*(z) = z^{k-1} - \frac{1}{d_n} A_n(z) - B_n(z) - d_n z C_n(z).$$

In order for $A_{n+1}(z)$ to be analytic at $z = 0$, we must have

(2.32)
$$d_n = \begin{cases} \dfrac{A_n(0)}{1 - B_n(0)} & \text{if } k = 1 \\ \dfrac{-A_n(0)}{B_n(0)} & \text{if } k > 1. \end{cases}$$

Determine δ_n as follows,

(2.33)
$$\delta_n = \begin{cases} 0 & \text{if } A^*_{n+1}(0) \neq 0 \text{ or } A^*_{n+1}(z) \equiv 0 \\ 1 & \text{if } A^*_{n+1}(0) = 0, \text{but } A^*_{n+1}(z) \neq 0. \end{cases}$$

The process terminates if $A^*_{n+1}(z) \equiv 0$ yielding a finite δ- fraction. Otherwise, the process continues. Also, note that $A_{n+1}(0) \neq 0$ and

(2.34)
$$B_{n+1}(0) = C_{n+1}(0) = \begin{cases} B_n(0) - 1 & \text{if } k = 1 \\ B_0(0) & \text{if } k > 1. \end{cases}$$

It is not hard to see that if $\delta_n = 1$ then $d_{n+1} = 1$ and hence the δ- fraction generated is a regular δ- fraction. ∎

The following is a summary of the algorithm for the nonsingular case. The algorithm for the singular case is analogous with only minor modifications required at the beginning.

Specify the number n of terms of the continued fraction to compute

Enter A_0, B_0 and C_0

If $A_0 \equiv 0$, let $\delta_0 = 0$ and stop

Determine δ_0, A_1, B_1 and C_1 by (2.9) and (2.10)

Determine d_1 by (2.14)

Determine A_2^*

If $A_2^* \equiv 0$, let $\delta_1 = 0$ and stop

Determine δ_1, A_2, B_2 and C_2 by (2.13) and (2.15)

Do $j = 3, \ldots, n + 1$

   Determine d_{j-1} by (2.21)

   Determine A_j^* by (2.20)

   If $A_j^* \equiv 0$, let $\delta_{j-1} = 0$ and stop

   Determine δ_{j-1}, A_j, B_j and C_j by (2.19) and (2.22)

3. Theory of the δ- fraction Solutions

In section 2 an algorithm was given for constructing a regular δ- fraction from a Riccati equation. The purpose of this section is to pursue the theory of the δ- fraction solution. First we point out that the regular δ- fraction resulting from the algorithm developed in the previous section is a formal continued fraction solution to the Riccati equation to which the algorithm was applied. The objective of the next theorem is to show that any formal δ- fraction solution to an admissible Riccati equation corresponds to the unique fps solution that vanishes at $z = 0$. A corollary to that theorem is that the formal δ- fraction solution guaranteed by Theorems 5 and 6 is the unique formal δ- fraction solution that vanishes at $z = 0$. The penultimate theorem states that an admissible Riccati equation has finite δ- fraction solution if and only if the Riccati equation has a rational solution. Finally, we close the section with a theorem that states the connection between a δ- fraction solution and an analytic solution.

Theorem 6. Let $R_0[W_0(z)] = 0$ be an admissible Riccati equation. The regular δ- fraction constructed by the algorithm given in the proof of Theorem 5 is a formal continued fraction solution to the Riccati equation.

Proof. We will prove it for the admissible Riccati equations of the form (2.1). The proof of the other case is completely analogous. From equations (2.7), (2.11), and (2.17) we have

$$
(3.1) \quad
\begin{cases}
R_0[W_0(z)] = R_1[W_1(z)] = \dfrac{-d_1}{(1 - \delta_1 z + W_2(z))^2} R_2[W_2(z)] \\[2ex]
\qquad = \displaystyle\prod_{k=1}^{n} \left(\dfrac{-d_k}{(1 - \delta_k z + W_{k+1})^2} \right) z^{n-1} R_{n+1}[W_{n+1}(z)].
\end{cases}
$$

Also, $W_0(z)$ is related to $W_1(z), W_2(z), \ldots$ by

$$
(3.2) \qquad\qquad W_0(z) = -\delta_0 z + W_1(z)
$$

and

$$
(3.3) \qquad W_0(z) = -\delta_0 z + \dfrac{d_1 z}{1 - \delta_1 z +} \; \cdots \; + \dfrac{d_n z}{1 - \delta_n z + W_{n+1}(z)}, \qquad \text{for } n \geq 2
$$

so setting $W_{n+1}(z) = 0$ determines $W_0(z)$. In fact, setting $W_{n+1}(z) = 0$ forces $W_0(z) = f_n$ where f_n is the nth approximant of the δ- fraction solution. Using this result with (3.1) we have

$$
(3.4) \qquad\qquad R_0[f_0(z)] = \dfrac{d_1}{(1 - \delta_1 z)^2} A_1(z)
$$

and

$$(3.5) \qquad R_0[f_n(z)] = \prod_{k=1}^{n-1} \left(\frac{d_k}{(1 - \delta_k z + W_{k+1}(z))^2} \right) \frac{d_n z^n}{1 - \delta_n z} A_{n+1}(z).$$

Since $\Lambda_0[W_k(z)] = O(z)$ we have $\Lambda_0 \left[\dfrac{d_k}{1 - \delta_k z + W_{k+1}(z)} \right] = O(1)$ and hence

$$(3.6) \qquad \Lambda_0[R_0[f_n(z)]] = O(z^n).$$

Therefore, by Definition 1, the δ- fraction solution given by the algorithm in Theorem 5 is a formal continued fraction solution to an admissible Riccati equation of the form (2.1). ∎

The purpose of the next theorem is to establish the connection between the formal power series solution and a formal δ- fraction solution.

Theorem 7. If $R_0[W_0(z)] = 0$ is an admissible Riccati equation, then a formal δ- fraction solution corresponds to the unique fps solution that vanishes at $z = 0$.

Proof. We prove the result for Riccati equations of the form (2.1). The proof is completely analogous in the other case. It is easy to establish the assertion that an admissible Riccati equation of the form (2.1) has a unique formal power series solution that vanishes at $z = 0$. From Theorem 2 we see that the relationship between a δ- fraction and its corresponding power series , $L(z)$, can be characterized by the equation

$$(3.7) \qquad L(z) - \Lambda_0(f_n(z)) = O(z^{n+1}).$$

Thus,

$$(3.8) \qquad \left\{ \begin{aligned} R_0[L(z)] &= R_0[(L(z) - f_n(z)) + f_n(z)] \\ &= R_0[f_n(z)] + (L(z) - f_n(z))B_0(z) + (L^2(z) - f_n^2(z))C_0(z) \\ &\quad -(L'(z) - f_n'(z)) \\ &= O(z^n) + O(z^{n+1}) + O(z^{n+1}) - O(z^n) \\ &= O(z^n) \end{aligned} \right.$$

(where the derivative of $L(z)$ is taken formally). This holds for every $n \in \mathbf{Z}^+$ and hence $L(z)$ must be the unique fps solution that vanishes at $z = 0$. ∎

Corollary 8. *Let $R[W(z)] = 0$ be an admissible Riccati equation. The δ- fraction solution guaranteed by Theorem 5 is the unique formal δ- fraction solution to the initial value problem*

$$(3.9) \qquad \begin{cases} R[W(z)] = 0 \\ W(0) = 0. \end{cases}$$

Proof. This is an immediate consequence of the uniqueness of the fps solution and Theorem 2. ∎

A consequence of this corollary is given in the following corollary.

Corollary 9. *An admissible Riccati equation has a rational solution that vanishes at $z = 0$ if and only if its formal δ- fraction solution is a terminating continued fraction.*

Proof. This result is easily obtained from Corollary 8 and Theorem 3. ∎

Our next theorem provides the connection between the formal δ- fraction solution and the unique analytic solution (if an analytic solution exists) that vanishes at $z = 0$.

Theorem 10. *Let $R[W(z)] = 0$ be an admissible Riccati equation. If the formal δ- fraction solution converges uniformly in a neighborhood of $z = 0$ to a function $W(z)$, then $W(z)$ is the unique solution of $R[W(z)] = 0$ that is analytic at $z = 0$ satisfying $W(0) = 0$.*

Proof. Let f_n be the nth approximant of the δ- fraction solution. By Theorem 5.13 in [7], $W(z) = \lim_{n \to \infty} f_n(z)$ is analytic in a neighborhood of $z = 0$, and the Taylor series expansion of $W(z)$ is the power series $L(z)$ to which the δ- fraction corresponds at $z = 0$. By Theorem 7, $L(z)$ is a fps solution of $R[W(z)] = 0$ at $z = 0$. It is therefore a solution of the Riccati equation in the neighborhood of $z = 0$ in which it converges. The assertion follows from the fact that $W(z) = L(z)$ for z in this neighborhood. ∎

4. Comments on Computational Aspects

One motivation for considering δ- fraction solutions is the desire for a more computationally satisfying algorithm than the C- fraction algorithm in [3,14]. A C- fraction is a finite or infinite continued fraction of the form

(4.1)
$$\frac{d_1 z^{e_1}}{1} + \frac{d_2 z^{e_2}}{1} + \cdots + \frac{d_n z^{e_n}}{1} + \cdots$$

where $d_n \in \mathbf{C}\setminus\{0\}$ and $e_n \in \mathbf{Z}^+$. A C- fraction is said to be a *regular* C- fraction if $e_n = 1$ for $n \in \mathbf{Z}^+$. The following two theorems from [3] are included to facilitate the comparison between δ- fraction solutions and C- fraction solutions.

Theorem 11. *(A) If $R[W(z)] = 0$ is an admissible Riccati equation of the form*

(4.2)
$$R[W(z)] := A(z) + B(z)W(z) + C(z)W^2(z) - W'(z) = 0$$

that has a regular C- fraction solution, then d_n is a function of the first n coefficients of $A(z)$ and $B(z)$ and the first $n-1$ coefficients of $C(z)$.

(B) If $R[W(z)] = 0$ is an admissible Riccati equation of the form

(4.3)
$$R[W(z)] := zA(z) + B(z)W(z) + C(z)W^2(z) - z^k W'(z) = 0$$

that has a regular C- fraction solution, then d_n is a function of the first n coefficients of $A(z)$, the first $n-1$ coefficients of $B(z)$ and the first $n-2$ coefficients of $C(z)$.

One significant consequence of this is that when computing, one may approximate the coefficient functions by polynomials without introducing any error into the computations due to these approximations. In other words, in order to calculate accurately the first n elements of the continued fraction, all one needs to use are the $(n-1)^{st}$ degree Taylor polynomials for $A(z)$ and $B(z)$ and the $(n-2)^{nd}$ degree Taylor polynomial (at $z = 0$) for $C(z)$.

Another consequence is that one can considerably reduce the computing by calculating fewer terms of the coefficient functions at each step. In order to calculate $d_1, \ldots, d_n$, we start with the $(n-1)^{st}$ degree Taylor polynomials for $A(z)$ and $B(z)$ and the $(n-2)^{nd}$ degree Taylor polynomials for $C(z)$. At each step, the degree of the polynomials is reduced by one unit until at the last step just the constant terms of $A(z)$ and $B(z)$ are calculated. These two terms are all that are needed to calculate d_n. Thus, the program is quite efficient.

When the C- fraction solution to the Riccati equation is not a regular C- fraction the computational story is more complex as is illustrated by the next theorem.

Theorem 12. *(A) If $R[W(z)] = 0$ is an admissible Riccati equation of the form (4.2) then the equation will have a C- fraction solution. If we define*

$$(4.4) \qquad\qquad \beta_n := \sum_{j=1}^{n} e_j \text{ and } \gamma_n := \sum_{j=2}^{n} e_j$$

then d_n is a function of the first β_n coefficients of $A(z)$, the first $\gamma_n + 1$ coefficients of $B(z)$ and the first $\gamma_n + 1 - \beta_1$ coefficients of $C(z)$.

(B) If $R[W(z)] = 0$ is an admissible Riccati equation of the form (4.3), then the equation has a C- fraction solution where d_n is a function of the first β_n coefficients of $A(z)$, the first γ_n coefficients of $B(z)$ and the first $\gamma_n - \delta_1$ coefficients of $C(z)$.

From this theorem, it is clear that if the C- fraction is not a regular C- fraction, then the numbers of coefficients needed to calculate d_n accurately are functions of the exponents $e_1, e_2, \ldots, e_n$. Therefore, one cannot be sure of the degrees of the polynomials to use in approximating the coefficient functions $A(z)$, $B(z)$ and $C(z)$ without a priori knowledge of the exponents in the continued fraction.

The δ- fraction solutions introduced in this paper have computational characteristics that are analogous to those in the case that the Riccati equation has a regular C- fraction solution. Before we proceed it is worth noting that every function analytic at $z = 0$ has both a regular δ - fraction expansion and a C- fraction expansion. These are identical in the case of a *regular* C- fraction expansion. The δ - fraction would then have $\delta_n = 0$ for $n = 1, 2, \ldots$ and the C- fraction would have $e_n = 1$ for $n = 1, 2, \ldots$.

Theorem 13. *(A) If $R[W(z)] = 0$ is an admissible Riccati equation of the form (4.2), then in the δ- fraction solution, δ_0 is a function of the constant term in $A(z)$, and for $n \in \mathbf{Z}^+$ d_n is a function of the first n coefficients in $A(z)$, the first $n - 1$ coefficients in $B(z)$, the first $n - 2$ coefficients in $C(z)$, and δ_n is a function of the first $n + 1$ coefficients in $A(z)$, the first n coefficients in $B(z)$, and the first $n - 1$ coefficients in $C(z)$. (If the number of coefficients is negative, then we assume there is no dependance on the coefficients.)*

(B) If $R[W(z)] = 0$ is an admissible Riccati equation of the form (4.3), then in the δ- fraction solution, δ_0 is a function of the constant term in $A(z)$, and for $n \in \mathbf{Z}^+$ d_n is a function of the first n coefficients in $A(z)$, $B(z)$, and $C(z)$, and δ_n is a function of the first

$n+1$ coefficients in $A(z)$, $B(z)$, and $C(z)$. (If the number of coefficients is negative, then we assume there is no dependance on the coefficients.)

Proof. (A) Let

(4.5)
$$\begin{cases} A_n(z) = a_{0,n} + a_{1,n}z + \ldots + a_{k,n}z^k + \ldots, \\[2mm] B_n(z) = b_{0,n} + b_{1,n}z + \ldots + b_{k,n}z^k + \ldots, \quad \text{and} \\[2mm] C_n(z) = c_{0,n} + c_{1,n}z + \ldots + c_{k,n}z^k + \ldots \ . \end{cases}$$

An inductive argument gives us that

(i) $a_{k,n}$ is a function of $a_{0,0}, \ldots, a_{k+n-1,0}, b_{0,0}, \ldots, b_{k+n-2,0}$ and $c_{0,0}, \ldots, c_{k+n-3,0}$ for $k = 0, 1, \ldots, n = 1, 2, \ldots$. (If $k+n-2 < 0$ or $k+n-3 < 0$, then there is no dependence on the coefficients of $B(z)$ or $C(z)$, respectively.)

(ii) $b_{k,n}$ and $c_{k,n}$ are functions of $a_{0,0}, \ldots, a_{k+n-2,0}$, $b_{0,0}, \ldots, b_{k+n-3,0}$ and $c_{0,0}, \ldots, c_{k+n-4,0}$ for $k = 0, 1, \ldots, n = 1, 2, \ldots$. (If $k + n - 2 < 0$, $k + n - 3 < 0$ or $k + n - 4 < 0$, then there is no dependence on the coefficients of $A(z), B(z)$ or $C(z)$, respectively.) The result now follows from the definitions of $\delta_n, n = 0, 1, \ldots$ and $d_n, n = 1, 2, \ldots$ given in the algorithm appearing in the proof of Theorem 5.

(B) The proof is analogous. ∎

As a result of this theorem, it is easy to see that the δ- fraction solutions have the same computational bonuses that regular C- fraction solutions have. The coefficient functions can be approximated by polynomials of appropriate degrees without introducing error into the computations up to a given n and the computations can be made very efficient (both as in the discussion following Theorem 11.) Thus, in the case when a Riccati equation does not have a regular C- fraction solution we can still enjoy the computational advantages associated with them if we turn to δ- fraction solutions instead of C- fraction solutions. Another advantage of using δ- fractions is that many convergence results are available for δ- fractions [11], whereas with C- fractions, virtually all of the convergence results apply to the special class of C-fractions, the regular C- fractions.

In closing, one more observation should be made. To the best of my knowledge, to date all of the δ- fraction expansions of functions have been obtained from known expansions in other types of continued fractions. This algorithm gives a method for obtaining δ- fraction expansions independent of other known expansions and independent of power series expansions. Clearly, the usefulness in this regard is restricted to functions that are solutions to admissible Riccati equations.

References

[1] RICHARD BELLMAN, ROBERT KALABA AND G. MILTON WING, *Invariant Imbedding and Mathematical Physics. I. Partical Processes*, J. Math. Phys, Vol.1, No.4, 1964, 280- 308.

[2] J. S. R. CHISOLM, *Continued fraction solution of the general Riccati equation*, Rational Approximation and Interpolation, Proc. of the UK- US Conf., Tampa, FL, 1983, Lecture Notes in Mathematics 1105 (Springer- Verlag, Berlin, 1984), 109- 116.

[3] K. D. COOPER, S. CLEMENT COOPER, AND WILLIAM B. JONES, *More on C-fraction solutions to Riccati equations*, to appear.

[4] S. CLEMENT COOPER, WILLIAM B. JONES, AND ARNE MAGNUS, *General T-fraction solutions to Riccati differential equations*, A. Cuyt(ed.), Nonlinear Numerical Methods and Rational Approximation, D. Reidel Publ. Co., 1988, 409- 425.

[5] S. CLEMENT COOPER, *General T- fraction solutions to Riccati differential equations*, Ph.D. dissertation, Colorado State University, 1988.

[6] WYMAN FAIR, *Padé approximation to the solution of the Riccati equation*, Math. of Comp. 18, 1964, 627- 634.

[7] WILLIAM B. JONES AND W. J. THRON, *Continued Fractions: Analytic Theory and Applications, Encyclopedia of Mathematics and Its Applications 11*, Addison- Wesley Publ. Co., Reading, MA, 1980, (distributed now by Cambridge Univ. Press, NY).

[8] J. KERGOMARD, *Continued fraction solution of the Riccati equation: Applications to acoustic horns and layered- inhomogeneous media, with equivalent electrical circuits*, to appear in Wave Motion.

[9] ALAN J. LAMB, *Algebraic aspects of generalized eigenvalue problems for solving Riccati equations*, C. F. Byrnes and A. Lindquist (eds.), Computational and Combinatorial Methods in Systems Theory, Elsevier Science Publ. B. V., North Holland, 1986, 213- 227.

[10] L. J. LANGE, *δ- fraction expansions of analytic functions*, Analytic Theory of Continued Fractions, Proc., Loen, Norway 1981, Lecture Notes in Mathematics 932 (Springer- Verlag, Berlin 1982), 152- 175.

[11] L. J. LANGE, *δ- fraction expansions for analytic functions*, SIAM J. Math. Anal., Vol. 14, No. 2, March 1983, 323- 368.

[12] G. C. MCVITTIE, *The mass- partical in an expanding universe*, Mon. Not. Roy. Ast. Soc. 93, 1933.

[13] G. C. MCVITTIE, *Elliptic functions in spherically symmetric solutions of Einstein's equations*, Ann. Inst. Henri Poincaré, Vol. 40, No. 3, 1984.

[14] E. P. MERKES AND W. T. SCOTT, *Continued fraction solutions of the Riccati equation*, J. Math. Anal. Appl. 4, 1962, 309- 327.

[15] A. N. STOKES, *Continued fraction solutions of the Riccati equation*, Bull. Austral. Math. Soc. 25, 1982, 207- 214.

IRRATIONAL CONTINUED FRACTIONS.

R.M. Hovstad.

Abstract. Irrational continued fractions are studied systematically by the aid of tails. The theory exposed extends the classical theory on irrational continued fractions.

Introduction. Irrationality studies of continued fractions go back to classical times. For example Lambert proved in 1761-1770 the irrationality of π by considering a continued fraction expansion of $\tan x$. Later Legendre proved in 1806 a general irrationality test. Stern contributed to this theory with a test in 1832. Also Stolz in 1886 and Pringsheim in 1908 have written on the subject. All these results are covered by the general classical irrationality test in [1] by Tietze in 1911 which essentially deals with continued fractions of the form

$$(1) \qquad \mathop{K}_{k=1}^{\infty} \frac{a_k}{b_k}$$

where $b_k > 0$ and $a_k \neq 0$ are integers for $k \geq 1$ obeying $b_k \geq |a_k|$ for $k \geq 1$ and even $b_k \geq |a_k|+1$ in case $a_{k+1} < 0$ for $k \geq 1$. It is concluded by Tietze that (1) converges to an irrational number if not $a_k < 0$ and $b_k = |a_k|+1$ for sufficiently large k. In this exceptional case (1) converges to a rational number. For all these developments see Perron's book [2, pp. 56-57]. As a general background to continued fractions, the recent book [3] is recomended.

After Tietze's general result in 1911 some authors have continued the study of irrational continued fractions along the same line as earlier. For example Bernstein and Szász studied the special case where essentially all elements of (1) are positive integers. They gave a fairly general result in this case. See their paper [4]. Later Fujiwara in [5] continued the studies of Bernstein and Szász and gave extremely general tests in the case where the elements in (1) are essentially positive integers. Fujiwara's results seem so far to be most advanced.

A different approach to irrationality studies was given in [6]. Here the elements in (1) are also positive integers. The technique used in [6] was based on continued fraction tails, which has been in focus by numerous authors in various connections recently, since the paper [8] of Waadeland in 1984. See for example [9].

A further exploration of the force of the method in [6] shows that more general results can be obtained where the signs of the partial numerators are not restricted to positivity. The method consists of the use of three term recurrence relations connected to tails. Consider the continued fraction

$$(2) \qquad \mathop{K}_{k=1}^{\infty} \frac{a_k}{b_k}$$

where $b_k > 0$ and $a_k \neq 0$ are integers for $k \geq 1$. The continued fractions

(3)
$$T_k = \underset{i=1}{\overset{\infty}{K}} \frac{a_i}{b_i}$$

are called the tails of (2) for $k \geq 1$. Suppose that (3) converges for all $k \geq 1$ to finite numbers. From (3) we conclude then that

(4)
$$T_k = \frac{a_k}{b_k + T_{k+1}}$$

is well defined and $T_k \neq 0$ for $k \geq 1$. Choose an integer $x_0 \neq 0$. Define

(5)
$$x_k = T_k x_{k-1}$$

for $k \geq 1$. From (4) and (5) we obtain the three term recurrence relation

(6)
$$a_{k-1} x_{k-2} = b_{k-1} x_{k-1} + x_k$$

for $k \geq 2$. The idea of the method is to combine (5) and (6). The equality (5) gives us the magnitude and sign of $x_k \neq 0$ and the equality (6) gives us that x_k is an integer if x_0 and x_1 are integers. The aim of an irrationality proof is then to prove the impossibility of such a sequence $\{x_k\}_{k \geq 0}$ in certain situations. In the opposite situations the aim is to establish rationality directly. The basis of an irrationality proof will therefore be the study of the sequence $\{x_k\}_{k \geq 0}$. More precisely we want to have a descending sequence of positive integers of the following type

(7)
$$\ldots < |x_{n_k}| < \ldots < |x_{n_1}| < |x_{n_0}| = |x_0|$$

where $\{n_k\}_{n \geq 0}$ with $n_0 = 0$ is a sequence of integers obeying $n_{k+1} > n_k$ for $k \geq 0$. It is clear that (7) gives us the wanted contradiction.

Descending sequences of positive integers in irrationality studies are not new. The earlier continued fraction studies of irrationality all base their methods on such sequences. The difference between earlier approaches and the present approach is that in the present approach such sequences are exploited differently. Tietze's classical result in [2, p. 56] is based on the choice $n_k = k$ for $k \geq 0$. A comment on Tietze's test in the light of (7) is given in [6].

In the next paragraph we give a general test which on the basis of the properties imposed on (1) gives necessary and sufficient conditions for (1) to be rational. Here the partial numerators are not restricted to positivity. The test gives therefore complete information. This test is based on the choice $n_k = 2k$ for $k \geq 0$ in (7). Numerous general tests can be based on the study of (7). They can all be seen as extensions of results given by Fujiwara in [5]. One of Fujiwara's results is given in a corollary to our test in the last paragraph containing applications. The difference between Fujiwara's approach and our approach lies in a more precise analysis of (7), in spite of Fujiwara's general lines of thought. The case studied by Bernstein and Szász is a simple special case of the general result by Fujiwara.

Before we go further it should be mentioned that the method described above can be adapted to the study of irrational functions represented by continued fractions. One notable example of this study is the recent treatment of the classical irrationality problem for T-fractions. See the paper [7].

The test. We will prove the following theorem.

Theorem (The test). Let the continued fraction

$$(8) \qquad \overset{\infty}{\underset{k=1}{K}} \frac{a_k}{b_k}$$

be given where $b_k > 0$ and $a_k \neq 0$ are integers for $k \geq 1$. We suppose that (8) with all its tails converge to finite numbers. In addition we suppose that

$$(9) \qquad b_{2k-1}b_{2k}b_{2k+1} + b_{2k-1}(a_{2k+1} - 1) \geq b_{2k+1}(|a_{2k-1}a_{2k}| - a_{2k})$$

for $k \geq 1$. Then (8) is a rational number if and only if

$$(10) \qquad a_{2k-1}a_{2k} > 0$$

and

$$(11) \qquad b_{2k-1}b_{2k}b_{2k+1} + b_{2k-1}(a_{2k+1} - 1) = b_{2k+1}(a_{2k-1}a_{2k} - a_{2k})$$

for k sufficiently large.

Proof: Define

$$T_k(n) = \overset{n}{\underset{i=k}{K}} \frac{a_i}{b_i}$$

for $n \geq k \geq 1$. If $x > \dfrac{a_{2k+1} - 1}{b_{2k+1}}$ for $k \geq 1$ we find that

$$(12) \qquad \frac{a_{2k-1} - 1}{b_{2k-1}} < \frac{a_{2k-1}}{b_{2k-1} + \dfrac{a_{2k}}{b_{2k} + x}} < \frac{a_{2k-1} + 1}{b_{2k-1}}$$

for $k \geq 1$. Putting $x = \dfrac{a_{2k+1}}{b_{2k+1}}$ for $k \geq 1$ and using (12) repeatedly we find that

$$(13) \qquad \frac{a_{2k-1} - 1}{b_{2k-1}} < T_{2k-1}(2n-1) < \frac{a_{2k-1} + 1}{b_{2k-1}}$$

for $n \geq k \geq 1$. In proving (13) we have used the socalled backward recurrence algorithm (see [3, p. 26]). Letting $n \to \infty$ we find from (13) that

(14)
$$\frac{a_{2k-1} - 1}{b_{2k-1}} \leq T_{2k-1} \leq \frac{a_{2k-1} + 1}{b_{2k-1}}$$

for $k \geq 1$. The double inequality (14) will be the basis of the proof.

Suppose that (8) is rational. Then

$$T_1 = \frac{x_1}{x_0}$$

for suitable integers $x_0 \neq 0$ and $x_1 \neq 0$. The identity (5) gives us $x_k \neq 0$ for $k \geq 0$. If we have strict inequality on both sides in (14) we obtain using (4) and (5) that

(15)
$$|x_{2k}| = |T_{2k}||x_{2k-1}| = |T_{2k-1}T_{2k}||x_{2k-2}|$$
$$= |a_{2k-1} - b_{2k-1}T_{2k-1}||x_{2k-2}| < |x_{2k-2}|$$

for $k \geq 1$. From (15) we see that we have a descending sequence of positive integers which is impossible. Thus we have equality in (14) at the left side or the right side for some $k \geq 1$. Suppose that

(16)
$$T_{2k-1} = \frac{a_{2k-1} + 1}{b_{2k-1}}$$

for some fixed $k \geq 1$. We know from (14) that $T_{2k+1} \geq \frac{a_{2k+1} - 1}{b_{2k+1}}$ for this particular k. Suppose that $T_{2k+1} > \frac{a_{2k+1} - 1}{b_{2k+1}}$. Putting $x = T_{2k+1}$ in (12) we find that $T_{2k-1} < \frac{a_{2k-1} + 1}{b_{2k-1}}$ which contradicts (16). Thus $T_{2k+1} = \frac{a_{2k+1} - 1}{b_{2k+1}}$ for this value of k. We can therefore conclude that in case of rationality we have equality at the left side of (14) for some $k \geq 1$. Therefore $T_{2k-1} = \frac{a_{2k-1} - 1}{b_{2k-1}}$ and

(17)
$$\frac{a_{2k-1} - 1}{b_{2k-1}} = \frac{a_{2k-1}}{b_{2k-1} + \dfrac{a_{2k}}{b_{2k} + T_{2k+1}}}$$

for some fixed $k \geq 1$. Again from (14) we know that T_{2k+1}

$$\geq \frac{a_{2k+1} - 1}{b_{2k+1}}$$ for this k. Suppose that $T_{2k+1} > \frac{a_{2k+1} - 1}{b_{2k+1}}$. Putting

$x = T_{2k+1}$ once more in (12) we find that $T_{2k-1} > \frac{a_{2k-1} - 1}{b_{2k-1}}$ which

contradicts (17). Thus $T_{2k+1} = \frac{a_{2k+1} - 1}{b_{2k+1}}$ for this value of k.
Doing the same to T_{2k+1} as we did to T_{2k-1} we can continue and
we conclude that in case of rationality we have

(18)
$$T_{2k-1} = \frac{a_{2k-1} - 1}{b_{2k-1}}$$

for k sufficiently large. From (18) we see that

(19)
$$\frac{a_{2k-1} - 1}{b_{2k-1}} = \cfrac{a_{2k-1}}{b_{2k-1} + \cfrac{a_{2k}}{b_{2k} + \cfrac{a_{2k+1} - 1}{b_{2k+1}}}}$$

for k sufficiently large. From (19) we obtain

(20)
$$b_{2k} + \frac{a_{2k+1} - 1}{b_{2k+1}} = \frac{a_{2k-1}a_{2k} - a_{2k}}{b_{2k-i}}$$

for k sufficiently large. Remembering (9) we conclude from (20)
that $a_{2k-1}a_{2k} > 0$ for k sufficiently large and conditions (10)
and (11) in the theorem are obtained.
On the other hand suppose that (10) and (11) in the theorem
are fulfilled. Then we notice that

(21)
$$\cfrac{a_{2k-1}}{b_{2k-1} + \cfrac{a_{2k}}{b_{2k} + \cfrac{a_{2k+1} - 1}{b_{2k+1}} + \cfrac{1}{x}}} = \frac{a_{2k-1} - 1}{b_{2k-1}} + \frac{1}{b_{2k-1} + a_{2k-1}a_{2k}x}$$

for sufficiently large k where x > 0. Repeated application of
(21) gives us

(22)
$$\frac{a_{2k-1}}{b_{2k-1}} + \ldots + \cfrac{a_{2n-1}}{b_{2n-1} + \cfrac{a_{2n}}{b_{2n} + \cfrac{a_{2n+1} - 1}{b_{2n+1}} + \cfrac{1}{x}}}$$

$$= \frac{a_{2k-1} - 1}{b_{2k-1}} + \cfrac{1}{b_{2k-1} + \sum_{i=k}^{n-1} b_{2i+1} \prod_{j=2k-1}^{2i} a_j + x \prod_{i=2k-1}^{2n} a_i}$$

for some $k \geq 1$, all $n \geq k+1$ and $x > 0$. Putting $x = b_{2n+1}$ and letting $n \to \infty$ in (22) we obtain

(23)
$$T_{2k-1} = \frac{a_{2k-1} - 1}{b_{2k-1}}$$

for some $k \geq 1$ since the last denominator in (22) tends to infinity (remembering (10) in the theorem). From (23) we conclude that (8) is rational after repeated application of (4) and the proof is completed.

Applications. In applications the convergence questions related to (8) and its tails are normally settled independently before we ask about irrationality. This is for example the case in Fujiwara's result (see [5]) in the corollary below.

Corollary (Fujiwara). Let the continued fraction

(24)
$$\mathop{K}_{k=1}^{\infty} \frac{a_k}{b_k}$$

be given where $b_k > 0$ and $a_k > 0$ are integers for $k \geq 1$. Suppose also that

(25)
$$b_{2k-1}b_{2k} + a_{2k} \geq a_{2k-1}a_{2k}$$

for $k \geq 1$. Then (24) converges to an irrational number.

Proof: Because of (25) the condition (9) in the theorem is fulfilled. Also convergence of (24) and all its tails to finite numbers is insured since (25) implies

(26)
$$\frac{b_{2k-1}b_{2k}}{a_{2k}} \geq a_{2k-1} - 1.$$

In fact by [2, Satz 2.11. p. 47] we conclude convergence if $a_{2k-1} \geq 2$ for infinitely many k. If on the other hand $a_{2k-1} = 1$ for k sufficiently large we have convergence trivially by the same result in [2]. Thus the theorem applies and we intend to conclude irrationality. Suppose that we have rationality of (24). Then from (11) and (25) we obtain $a_{2k+1} = 1$ for k sufficiently large. Thus we have $b_{2k-1}b_{2k} + a_{2k} = a_{2k-1}a_{2k}$ for k sufficiently large which is clearly impossible since $a_{2k-1} = 1$ and $b_{2k-1}b_{2k} \geq 1$. Therefore (24) is irrational and the corollary is proved.

As we see our result in the theorem is far more precise than Fujiwara's result in the corollary both with respect to arbitrary signs of the partial numerators and the complete information it gives about the possibility of rationality. We will give an example of rationality in the theorem where the elements are positive.

Example (Positive rational case). The continued fraction

(27)
$$p = \frac{p+1}{1} + \frac{3}{2p} + \frac{p+1}{1} + \ldots$$

where $p > 0$ is an integer converges with all its tails to finite numbers according to [2, Satz 2.11. p. 47]. Putting $a_{2k-1} = p+1$, $b_{2k-1} = 1$, $a_{2k} = 3$ and $b_{2k} = 2p$ for $k \geq 1$ in the theorem we find that (10) and (11) are fulfilled and we conclude rationality. The actual value can easily be found since (27) is a 2-periodic continued fraction.

It is interesting to note that Fujiwara's result in the corollary can not handle (27) and that Tietze's classical test does not include positive continued fractions in the rational case like in the present theorem.

References.

1. H. Tietze: Über Kriterien für Konvergenz und Irrationalität unendlicher Kettenbrüche, Math. Ann. 70, 1911.

2. O. Perron: Die Lehre von den Kettenbrüchen, Band II, 3. Aufl., 1957, Teubner, Stuttgart.

3. W.B. Jones and W.J. Thron: Continued Fractions, Analytic Theory and Applications, Encyclopedia of Mathematics and Its Applications, 1980, Addison-Wesley, Reading, Massachusetts.

4. F. Bernstein and O. Szász: Über Irrationalität unendlicher Kettenbrüche mit einer Anwendung auf die Reihe $\Sigma q^{\nu^2} x^\nu$, Math. Ann. 76, 1915, 295-300.

5. M. Fujiwara: Über Irrationalität unendlicher Kettenbrüche, Science Reports of the Tohoku Imperial University 8, 1919, 1-10.

6. R.M. Hovstad: Continued fraction tails and irrationality, Rocky Mountain J. Math. (to appear).

7. R.M. Hovstad: The classical irrationality problem for T-fractions, Proc. Amer. Math. Soc. (to appear).

8. H. Waadeland: Tales about tails, Proc. Amer. Math. Soc. 90 (1984), 57-64.

9. W.J. Thron (ed.): Analytic Theory of Continued Fractions II, Proceedings, Pitlochry and Aviemore, 1985, Lecture Notes in Mathematics, 1199, Springer-Verlag. Berlin, Heidelberg, 1986.

Julius Worpitzky, his contributions to the analytic theory of continued fractions and his times

Lisa Jacobsen
W. J. Thron
Haakon Waadeland

1. Introduction. Our story, in a way, begins in 1898 with the publication of an article on convergence of continued fractions with complex elements by A. Pringsheim, then an "auszerordentliche Professor" at the University of München. In it he has, among others, the result:

> *The continued fraction* $K(a_n/1)$, *all of whose elements* a_n *are complex numbers, converges if* $|a_n| \leq 1/4$ *for all* $n \geq 1$.

This theorem was obtained 33 years earlier by Julius Worpitzky, then a teacher at the Friedrichs-Gymnasium in Berlin. Pringsheim returned numerous times to this result without ever acknowledging Worpitzky's priority, even though he must have known about Worpitzky well before 1925 when he included a long chapter on continued fractions with elements in the complex plane in his book "Zahlen- und Funktionenlehre".

It is because of this, hard to understand, oversight, which was shared by many other mathematicians, that we became interested in finding out who Worpitzky was, how he proved his theorem, and why nobody appears to have known about it before 1905.

Pringsheim relied on Stolz for his belief that there had been no general convergence results for continued fractions with complex elements before 1886. In that year the latter in his book "Vorlesungen über allgemeine Arithmetik, zweiter Theil: Arithmetik der complexen Zahlen" has the following to say. "There are no results of any generality on convergence of continued fractions with complex elements except for the theorem: "$K(1/b_n)$ diverges if $\sum |b_n|$ converges and results for the convergence behavior of periodic continued fractions."

In 1905, or shortly before then, someone brought Worpitzky's article to the attention of Van Vleck (himself a rediscoverer of the theorem) who then acknowledged Worpitzky's priority in his "Selected Topics in the Theory of Divergent Series and of Continued Fractions" in the "Boston Colloquium". Van Vleck thought that the paper was Worpitzky's dissertation which makes it likely that he never saw the article itself. E. Wölffing in his 1908 article "Wer hat über Kettenbrüche gearbeitet", which is simply a list of publications on continued fractions, has a reference to Worpitzky. Perron in his generally careful and definitive book "Die Lehre von den Kettenbrüchen" of 1913 makes no mention of Worpitzky (the name finally appears in the third edition in 1957).

The next reference that we are aware of occurs in a footnote in a long article by O. Szász (Journal f.d.r.u.a. Math., vol. 147, p. 160) in 1917. It reads as follows:

> Finally, attention should be called to a noteworthy contribution by Julius Worpitzky (reference) which appears to be little known. (Compare also Van Vleck, The Boston Colloquium, etc.) Worpitzky proved (in §22) that the continued fraction $1/(1 + K(a_\nu x/1))$, where the a_ν satisfy $|a_\nu| \leq a$ $(\nu = 1,2,3, \cdots)$ converges for $|x| < 1/4a$ and represents a synectic (holomorphic) function. Later proofs of this theorem were given by Van Vleck and Pringsheim. Worpitzky's proof contains a small gap but this can be filled easily. From his arguments one can also conclude directly that the convergence is uniform.

Szász appears to have read Worpitzky's article. It seems possible that he was the only one to have done so until we studied it recently. Szász referred to Worpitzky at least once more. That was in an article presented to the "Sitzungsberichte der Münchner Akademie" by Pringsheim in 1919.

The first reference in a text on continued fractions is by Wall (a student of Van Vleck) in his "Analytic Theory of Continued Fractions" in 1948. He calls the result "Worpitzky's Theorem" (Chapter III, §10) and introduces it with the comment:

> In what appears to be the earliest published paper treating of the convergence of continued fractions with complex elements, Worpitzky showed that $1/(1 + \overset{\infty}{\underset{\nu=2}{K}} (a_\nu/1))$ converges if the partial numerators $a_2, a_3, a_4, \cdots$ all have moduli less than $1/4$.

Stolz appears to have given a fairly correct picture of the state of development of convergence theory of continued fractions in 1886. However, in addition to having overlooked Worpitzky, he also did not mention a result of Oppermann which was published posthumously by J.P. Gram in 1883. The theorem is: $K(1/(b_n i^{\alpha_n}))$ *converges if* $b_n \geq 2$ *for all* n. No proof was given nor was the nature of the α_n specified.

Immediately after Stolz's assessment in 1886 a spate of new results was obtained in many parts of the world. The first results were due to Sleshinskii in Russia, who in 1888 and 1889 proved among others that $K(a_n/b_n)$ converges if $|b_n| \geq |a_n| + 1$ and that $K(a_n/1)$ converges if $|a_n| \leq 1/4$. He was followed by Pincherle in Italy in 1889 who had the result that $K(1/b_n)$ converges if $|b_n| \geq 2 + \epsilon$ (this is contained in one of Sleshinskii's results). In 1895 von Koch in Sweden rediscovered another one of Sleshinskii's criteria, namely: $K(a_n/1)$ converges if $\sum |a_v| < \infty$. Both authors noted that numerators and denominators of the approximants approach separate limits. Then came the discoveries by Pringsheim in Germany between 1898 and 1910 and Van Vleck in U.S.A. from 1901 to 1904, both had a substantial overlap with earlier work.

Even today Worpitzky's theorem is one of the most useful convergence results for continued fractions with complex elements. It also has the distinction of being the first "convergence region" criterion. Since then many more results of this type, which in many respects is the most typical for convergence theorems for continued fractions, have been found.

2. Early Life. Julius Daniel Theodor Worpitzky was born on May 10, 1835 in the small village of Carlsburg (now Karlsburg) near Greifswald in Pomerania. His father Johann Samuel (1804-59) was a school teacher in Carlsburg. His grandfather Johann Christian Worpitzky (1775-1857) was a damask weaver in Carlsburg as was one of his uncles (1802-56, committed suicide). Another uncle became a teacher at the Gymnasium in Parchim. The third brother of his father was cook to the count at Carlsburg. Julius' mother was Frederike Henriette, her maiden name was Detloff. Otherwise we know nothing about her. Julius was the first living child of his parents. He also had a sister Carolina Maria Johanna (1839-60) who died of tuberculosis at the age of 21. Her father had died of the same disease the year before. One wonders whether weavers in Carlsburg had as hard a life as those in Silesia, who, in desperation, rose in revolt against their masters in 1844.

We know nothing about Julius' early schooling. Presumably it took place in his father's school in Carlsburg. In the fall (Michaelis) 1848 he was enrolled in the Gymnasium at Anclam (now Anklam). Though Carlsburg is located in Greifswald County (Kreis) it is only 12 km from Anclam, while it is 25 km from Greifswald. This may explain why Julius attended the Gymnasium in Anclam. Whether he was able to commute daily to Anclam or whether he stayed there during the week, possibly with relatives, is not known. He graduated (Zeugnis der Reife) Easter 1855 as the first in a class of 10 pupils. He then went on to the University of Greifswald where he studied mathematics and science until the fall of 1857. He completed his studies at the University of Berlin in the winter semester 1857/8.

The city of Greifswald was settled around 1240 by people coming from the Netherlands. It became a member of the Hanseatic League in 1278. The university was founded in 1456. From 1648 to 1815 Greifswald and the surrounding territory (Vorpommern) were under Swedish rule. In 1815 it came to Prussia.

During Worpitzky's stay at the University of Greifswald the Ordinarius in mathematics was Johann August Grunert (1797-1872) who succeeded Johann Karl Fischer (1761-1833) in '33 and was succeeded by Bernhard Minningerode (1837-1896) in '74. In addition Gustav Salomon Tillberg (1777-1859) was teaching mathematics (as an associate professor?) at the university. His successor was Leo Königsberger who came as an associate professor in '64 and was made a full professor in '66. Königsberger left Greifswald in '69; he was followed by Fuchs '69-'74 and Thomé '74-'10. Grunert appears to have been the one who had most influence on Worpitzky.

After studying in Halle and Göttingen Grunert received his doctor's degree from the University of Halle in 1820. He then taught at the Gymnasium in Torgau (1821-28) and Brandenburg (1828-33). Grunert was best known for founding the "Archiv der Mathematik und Physik" in 1841. He continued as editor thoughout his life. The main purpose of the journal was to establish a connection between school and university mathematics and to keep alive the enthusiasm of the secondary school teachers for science. Grunert completed, in part together with Mollweide, the "Mathematische Wörterbuch" of Klügel and issued two supplementary volumes. Grunert was a prolific publisher (about 500 titles) including at least two articles on the theory of continued fractions.

E. Lampe (Worpitzky's colleague at the Kriegsakademie and a professor at the Technische Hochschule in Berlin) wrote two, almost identical, obituaries of Worpitzky. In these he suggests that Grunert may have been out of touch with the latest mathematical developments when Worpitzky was his student at Greifswald. This had an effect on Worpitzky's mathematical development. Lampe suggests that it made Worpitzky more independent and possibly somewhat stubborn on the one hand, but inadequately familiar with the literature of the subjects he studied on the other hand. Cantor, the mathematical historian, also was critical of Grunert (in his biography of Grunert in the "Allgemeine Deutsche Biographie") while admitting the quantity of his output he was doubtful of the lasting importance of his work.

There may also have been a scarcity of able fellow students for Worpitzky to interact with and measure himself against at Greifswald in those days.

Worpitzky completed his studies in Berlin during the winter semester 1857/8. The University of Berlin (UB) was founded in 1810. While efforts to have outstanding faculty were successful in other fields, the initial appointments in mathematics appear to have been injudicious. Even though Dirichlet, Eisenstein and Steiner had been associated with it earlier, it was only when both Kummer and Weierstrass, the former as Ordinarius in 1855 and the latter as auszerordentlicher Professor in 1856, joined the department that the study of mathematics at UB was put on a solid foundation. During the semester that Worpitzky spent in Berlin, Kummer and Martin Ohm were the full professors (Ordinarius = ordentliche Professor). Weierstrass and Steiner were the associate professors (auszerordentliche Professoren) and Arndt, Borchardt and Hoppe were lecturers (Privatdozenten). Dirichlet had left to become Gauss' successor in Göttingen and Eisenstein had died in 1852. Unfortunately we do not know whose lectures Worpitzky attended, which subjects he studied and which of his fellow students he became acquainted with.

The next three years find Worpitzky first in 1858/9 as a tutor (Hauslehrer) in the Lausitz and then in the Russian province of Livland as a private teacher (Privatlehrer) until the summer of 1862. We know that in 1861 he was in Pernau (Pärnu) on the Gulf of Riga about 100 miles north of Riga in what now is the Estonian SSR. We can only speculate on what caused this long interruption in his professional career but it seems likely that finances had a lot to do with it. As it is, it is surprising that his father, on a village school master's wages, was willing and able to support his son through six years of Gymnasium and three years of university. However it appears that even in the first half of the 19th century the desire to see one's children move up in the world and one's ability to do something about it was already greater than one might have expected. It should also be noted that the study of mathematics and science carried considerably less prestige in those days than that of other academic disciplines such as law, medicine and philosophy.

The death of the father in 1859 had undoubtedly something to do with Worpitzky's need to earn and save money before he was able to complete his professional training.

3. First Mathematical Publication. By 1860 a good deal of work had been done by Gauss, Bolzano, Cauchy, Abel, Dirichlet and others to clarify the concepts of limit and convergence. However these "innovations" had not yet found general acceptance in the mathematical community. As late as 1885 Stolz in the preface to the first volume of his "Vorlesungen über allgemeine Arithmetik" speaks about the substantial divergence between the popular pedagogical and the scientific presentation of the elements of analysis.

A very good account of the pre-Cauchy state of affairs concerning convergence of series can be found in J.V. Grabiner's "The origins of Cauchy's rigorous Calculus" (pp. 99-101 and footnotes 75, 76). The two concepts: "nth term tends to zero" and "nth partial sums approach a limit" simply existed side by side. This is in part justified by the fact that for power series in the interior of their region of convergence the two concepts coincide.

The co-existence can still be found in G.S. Klügel's "Mathematisches Wörterbuch" (Leipzig 1803-31, completed by Mollweide and Grunert). Worpitzky was familiar with Klügel's Wörterbuch. He referred to it repeatedly in his article in Crelle's Journal in 1883. So it is not surprising to find in both of Worpitzky's articles on continued fractions (1862 and 1865) as a necessary and sufficient condition for the convergence of a continued fraction with approximants f_n the condition $f_n - f_{n-1} \to 0$.

Standards of rigor in mathematical arguments were increasing throughout the 19th century. But again this spread was not uniform. Grunert in particular may not have been much influenced by this trend. Thus (according to Sleshinskii, Odessa Proceedings 1889) a convergence proof for continued fractions with positive elements given by Grunert in 1838 was incorrect. The error was corrected by Arndt, a student of Grunert, in his Greifswald thesis in 1845. Thus it is not surprising that Worpitzky did not always demonstrate the degree of rigor one would expect today. After this lengthy preamble we turn to the main topic of this section.

Worpitzky's first mathematical article was printed in Greifswald in 1862. It was concerned with solutions of the Riccati differential equation

$$(3.1) \qquad\qquad y' + y^2 = ax^{m-2}$$

In addition to other types he considers solutions in terms of continued fractions of the form

$$(3.2) \qquad xy = 1 + \mathop{\mathrm{K}}_{n=1}^{\infty} \left(\frac{ax^m}{nm \pm 1} \right).$$

Lagrange in 1776 and Euler (posthumously in 1813/4) had both used continued fractions to solve differential equations. The latter had expressions very similar to (3.2). Worpitzky was unaware of this earlier work. The main merit (from our perspective) of Worpitzky's paper is that both the question of convergence of (3.2) as well as the question whether it converges to a solution of (3.1) are addressed.

In the preface to the printed version of the article, dated on his 26th birthday (1861) Worpitzky acknowledges letters from Kummer, Helmling and Grunert who had pointed out that the two former had given solutions (but not in terms of continued fractions) of (3.1) which he had not referred to. He corrected this oversight in the printed pamphlet. It seems likely that Worpitzky sent a manuscript of his results to Grunert and possibly the other two. If so, it is surprising that none of them mentioned the earlier work of Lagrange and Euler. One also wonders whether the errors which Worpitzky made in the paper were noticed.

Using modern terminology we can set

$$S_n(w) = 1 + \frac{ax^m}{m+1} = \cdots \frac{ax^m}{nm+1} + \frac{ax^m}{nm+n+w}.$$

Then the nth approximant $f_n = S_n(0)$. Worpitzky had established that

$$(3.3) \qquad xy = S_n(w_n),$$

where y is a solution of (3.1) and w_n is a solution of a related differential equation. It was clear to him that from (3.3) one cannot conclude without additional arguments (and assumptions) that

$$(3.4) \qquad xy = \lim_{n \to \infty} f_n = \lim_{n \to \infty} S_n(0).$$

He gives two "proofs" for the transition from (3.3) to (3.4). The first relies on the observation that $w_n \to 0$, which in itself is not sufficient to conclude that (3.4) holds. The second approach is by means of a general result to the effect that if one has a convergent continued fraction and $\{g_n\}$ is an arbitrary sequence then

(3.5)
$$\lim_{n\to\infty} S_n(g_n) = \lim_{n\to\infty} S_n(0) .$$

For the result to be true one needs the additional condition

(3.6)
$$|g_n + S_n^{-1}(\infty)| > \epsilon .$$

Worpitzky then proceeds to investigate when the continued fraction under consideration converges. As already mentioned he considers

(3.7)
$$\lim_{n\to\infty} (f_n - f_{n-1}) = 0$$

as necessary and sufficient for the convergence of his continued fraction (see p. 16, l 5 and p. 17, cond. 1). It happens that in the applications he has in mind $f_n - f_{n-1} = O(r^n)$, $0 < r < 1$, so that the condition (3.7) is sufficient for convergence of the continued fraction.

In another place (p. 16, l 12) Worpitzky states that $\lim r_n = 0$ and $r_n > 0$, $n \geq 1$ implies $r_{n+1} < r_n$ from some N on. Again, monotonicity is not required to prove his result.

To sum up: Worpitzky shows both ingenuity and insight. His major failing, beyond being a child of his time, appears to be that he likes to assert results in a generality, which makes them untrue, while the restricted statements that are sufficient to carry his argument, are likely to be true.

In the appendix to the article Worpitzky studies the expansions of e^x and $\tan x$, both solutions of certain Riccati equations. His interest in explicit numerical data such as truncation errors (pp. 66, 71) is noteworthy.

It should also be noted that all results are considered on the real axis only, though extension to the complex plane would have been feasible.

4. The Mathematical Environment in Berlin, 1862-95. In the fall of 1862 Worpitzky returned to Prussia and joined Schellbach's mathematical-pedagogical seminar at the Friedrich Wilhelm Gymnasium in Berlin. Worpitzky spent the rest of his life in Berlin so it may be of interest to give a short account of mathematical activity in Berlin during this period. We shall see that the number of people engaged in mathematical endeavors was still quite small and that there were many possible contact points between Worpitzky and others in this group.

The center of research was the University (UB) which during most of this time was dominated by Kummer, Weierstrass and Kronecker. (Biographical data are listed at the end of this section.) In addition Arndt was an associate professor until his death in '66. Borchardt was important both for his research and as editor of the "Journal für die reine und angewandte Mathematik" (he succeeded Crelle in '56). He also lectured at UB. Hoppe became a lecturer at UB in '53 and continued in this position until his death. He was an active researcher but a poor teacher and found little recognition (he finally was granted the title "professor" in '70).

During this time UB undoubtedly had the best mathematics department in Germany. Göttingen, in second place, probably became the better of the two after Hilbert had joined Klein there in 1895.

The next generation of professors at UB was to a large extent already educated in Berlin. Fuchs lectured at UB from '65-69 and returned as a full professor (Kummer's successor) in '84. Frobenius was an associate professor briefly in '74/75 and came back as Kronecker's successor in '92. H. A. Schwarz after studying at the TH and receiving his doctor at UB in '64 was called back after a distinguished career at ETH and Göttingen, as his teacher's (Weierstrass) successor in '92.

People who, after having reached the rank of associate professor at UB went on to become full professors at other institutions were Thomé, Bruns, Wangerin, Netto and Hensel. Knoblauch also worked his way up to the rank of associate professor at UB in '90 and stayed in this position until his death in 1915. Pochhammer, Runge, Kötter and Schlesinger left UB, each after having served as a lecturer for a few years. Finally, L. Königsberger in '60, Gordan in '62, Lampe in '64, G. Cantor in '67 and Schottky in '75 were among those who received doctorates from UB during this period.

There were three technical institutions in Berlin. The oldest one was the "Bergakademie"; it was followed in 1799 by the "Bauakademie" and finally in 1821 by the " Gewerbeakademie" (originally called "Technisches Institut"). The last two schools were combined into the "Technische Hochschule" (TH) in 1879. In 1916 the Bergakademie also became part of the TH.

The first to teach mathematics at the Technische Institut were Grashoff, an engineer, and Pohlke, an artist, who taught mainly geometry. Schellbach and possible other secondary school teachers also taught there for awhile. In '56 Weierstrass was appointed to the "1. Lehrstelle" with the title professor (and a salary of 1500 Thaler). He was succeeded by Aronhold, a respectable geometer, who held the position from '64 to '83. Then came H. Weber for a year. After him P. Du Bois-Reymond filled the position for five years. His successor in '89 was Lampe. A "2. Lehrstelle" was created for Christoffel in '69 (at a salary of 1800 Thaler). He was succeeded by Kossack in '72. Then Stahl held the position for a short time to be followed by Hettner, who was also an associate professor at UB. Buka was a lecturer in '79 and became a professor in '92. Weingarten taught at the Bauakademie starting in '66 as a lecturer. He became a professor in '71 and retired (by now from the TH) in '02.

Originally the TH was not equivalent to a university. This applied to entrance requirements and privileges of students as well as duties and privileges of faculty members. However, changes made between '60 and '99 resulted in the TH being equal to a university in all essential respects. But even before '99 mathematicians moved back and forth freely between these two types of institutions. Thus Weierstrass went from the Gewerbeakademie to a full professorship at UB in '64, Du Bois-Reymond had been an Ordinarius at Tübingen before coming to the TH. Christoffel went on to become a full professor at the newly founded University of Strassburg, where he was joined a few years later by H. Weber. Aronhold rejected calls from a number of universities. Joint appointments with the UB were relatively rare. We only know of Weierstrass and Hettner. However there were a number of teachers in secondary schools who also taught at the TH (usually as lecturers). We can name Müller, Servus and Buka. There was probably more emphasis on teaching ability at TH than there was at UB. But even at the latter teaching was important (recall the fate of Hoppe).

The "Kriegsakademie" in which young officers with at least three years of service were given a three year course to prepare them for higher staff positions, employed a number of mathematicians on a part-time basis. Dirichlet taught there from '28 until he left Berlin. Kummer gave courses between '55-'74. Schellbach from '43, Worpitzky from '72 and Lampe from '74 held "Lehraufträge" at the Kriegsakademie while their main employment was at secondary schools.

Many of the mathematicians who eventually became faculty members at UB, TH or other universities had taught school at the secondary level for a number of years (not infrequently more than ten). The list includes Grunert '20-'33, Kummer '31-'42, Arndt '40(?)-'54, Weierstrass '41-'55, Hertzer '55-

'65(?), Fuchs '58-'65, Weingarten '58(?)-'66(?), Du Bois-Reymond '59-'65, Königsberger '61-'64, Schwarz '64-'67(?), Lampe '64-'89, Kossack '65-'73, Wangerin '67-'76, G. Cantor '67-'69, Netto '70-?, Frobenius '72(?)-'74, Buka '77-'96, Hettner '77-'82(?).

The Deutsche Mathematiker Vereinigung (DMV) in its first membership list in 1891 has a total of 205 names. Worpitzky was one of the 33 living in Berlin. Ten of the 33 were connected with secondary schools and two were librarians.

Hoppe, whom we have already encountered as a Privatdozent at UB succeeded Grunert after the latter's death in '72 as editor of the "Archiv für Mathematik und Physik" and continued in this position until his death in 1900. We also have already mentioned that Borchardt took over from Crelle in '56 as editor of the "Journal für die reine und angewandte Mathematik". He was assisted by Schellbach, Kummer, Kronecker and Weierstrass (listed in that order). After Borchardt's death in '80 Kronecker and Weierstrass, jointly, became the editors of the journal. When Ohrtmann together with F. Müller and Wangerin started the "Jahrbuch über die Fortschritte der Mathematik" (the first reviews were for articles published in '68). Lampe became a contributor. He was co-editor with Henoch from '85-'90 and after '90 he was the sole editor of this important review journal until his death in 1918.

Schellbach's main position was as a professor at the Friedrich Wilhelm Gymnasium. He was also the director of the mathematical-pedagogical seminar from its beginning in '55 until it was laid down in '80. The purpose of this seminar was to prepare university graduates for teaching mathematics in the secondary schools. It was not the only gateway into the teaching profession. One could also start directly with a "Probejahr" in a school. Among the seminars many students, who usually remained Schellbach's friends after they had moved into important positions, were Clebsch, C. Neumann, Fuchs, Königsberger, H.A. Schwarz, F. Müller, G. Cantor, Weingarten and Worpitzky. Schellbach was a more important and influential man than his positions may make one believe. In addition to ties to his former students he also had extremely good connections both in the Ministry of Education and in the royal family. He had tutored the crown prince (later emperor Friedrich III) in the mathematical sciences. He was also instrumental in securing for mathematics and the sciences an equal position with the classical languages in the prussian gymnasia.

As in so many of his statements E.T. Bell was less interested in accuracy and more intent on creating an effect when he referred to Schellbach (without mentioning his name) as "an uninspired mathematical mediocrity" in the chapter on Cantor in "Men of Mathematics".

Mathematicians who were connected with UB **between** '63-'95 (degrees and appointments at UB. Dr = doctor, Pd = Privatdozent, aoP = auszerordentlicher Professor, oP = ordentlicher Professor, Ak = Mitglied der Akademie)

First Generation:
    Ohm (1792-1872, Pd 21, aoP 24-39, oP 39-68)
    Kummer (1810-93, oP 55-83)
    Weierstrass (1815-97, aoP 56-64, oP 64-92)
    Kronecker (1823-91, Dr 45, Ak 61, oP 83-91)
    Arndt (1817-66, Pd 53-64, aoP 64-66)
    Borchardt (1817-80, Pd 48-80, Ak 55)
    Hoppe (1816-1900, Pd 53-1900)

Second generation:
    Fuchs (1833-02, Dr 58, Pd 65-66, aoP 66-69, oP 84-02)
    Frobenius (1849-17, Dr 70, aoP 74-75, oP 92-16)
    Schwarz (1843-21, Dr 64, oP 92-17)

aoP:
    Thomé (1841-10, Dr 64, Pd 69-70, aoP 70-74)
    Bruns (1849-19, Dr 71, aoP 76-81)
    Wangerin (1844-33, aoP 76-82)
    Netto (1846-19, Dr 70, aoP 82-88)
    Hettner (1854-14, Dr 77, aoP 82-14)
    Knoblauch (1855-15, Dr. 82, Pd 83-90, aoP 90-15)
    Hensel (1861-41, Dr 84, Pd 86-92, aoP 92-02)

Pd:
    Pochhammer (1841-20, Dr 63, Pd 72-74)
    Runge (1856-27, Dr 80, Pd 83-86)
    Kötter (1859-22, Dr 84, Pd 87-97)
    Schlesinger (1864-33, Pd 89-97)

Mathematicians with appointments at TH:
    Grashoff (1826-93, before 69)
    Pohlke (1810-76, before 69)
    Schellbach (1805-92, before 69)
    Weierstrass (1915-97, oP 56-64)
    Aronhold (1819-84, oP 64-83)
    Christoffel (1829-1900; Dr 56 UB, Pd 59-62 UB, oP 69-72)
    Kossack (1839-92, Pd 69; Prof. 75-92)
    Hertzer (1831-08, 55, 65; Prof. 71, festeStelle 75-07)
    H. Weber (1842-13, oP 83-84)
    P. Du Bois-Reymond (1831-89, Dr 59 UB, oP 84-89)
    Lampe (1840-18, Dr 64 UB, oP 89-18)

W. Stahl (1846-94, oP 92-94)
Hettner (1854-14,   )
Buka (1852-96, Pd 79, Doz. 89, t.P 92; concurrently Real Gymnasium)
Weingarten (1836-10, Pd 66?, Prof. 71)

5. Worpitzky in Berlin (1862-95). When Worpitzky returned to Berlin in '62 he joined, as we have already mentioned, Schellbach's seminar from fall '62 to Easter '63. While there he helped with Schellbach's book "Elliptische Integrale und Thetafunctionen" which appeared in '64. Worpitzky's assistance is acknowledged in the preface of the book. Easter '63 he started his teaching career as a "Hülfslehrer" at the Friedrichs Gymnasium und Realschule and was promoted to the rank of ordinary teacher on Jan 1, 1864 after having passed the examination "pro facultate docendi" on June 2, 1863. His original assignment was to the "Mittelschule". He continued in this position until he was transferred to the Friedrichs-Werdersche Gymnasium in '68.

Worpitzky's continuing interest in scholarship and research was shown by his joining the "Physikalische Gesellschaft zu Berlin" in '63 and by attending its meetings faithfully at least in the earlier years of his membership. The organization was at that time the meeting place of all of Berlin's young physicists and mathematicians.

In '65 he published the article which aroused our interest in him: "Untersuchungen über die Entwickelung der monodromen und monogenen Functionen durch Kettenbrüche. Erste Folge" in the "Jahresbericht" of his school. In July '67 he received his doctorate for a thesis on the finiteness of definite integrals and sums of series from the University of Jena. In a later chapter we shall offer some conjectures trying to explain why he did not get his doctorate from Greifswald or Berlin which would appear to have been the natural choices. Jena at that time had Carl Snell (1806-79) as Ordinarius and Hermann Schaeffers as associate professor. Neither of the two was distinguished for his research. Snell, who as Ordinarius must have been the one to accept Worpitzky's dissertation, was mainly interested in pedagogical questions. Beyond listing Worpitzky as having received his doctorate in '67 there is no reference to him in the archives of the University in Jena. However there is a letter of recommendation from Schellbach for another candidate. It thus seems likely that it was Schellbach, himself the recipient of a doctorate from Jena, who suggested Jena to Worpitzky and, most likely, recommended him to Snell. Quite possibly the degree was conferred "in absentia".

When Worpitzky was transferred to the Friedrichs-Werdersche Gymnasium at Easter '68 he was assigned the third ordinary teaching position (3. ordentliche Lehrstelle). Others who had taught at this Gymnasium or the affiliated Gewerbeschule whose names we have encountered before, were Schellbach, Weingarten, Du Bois-Reymond, Lampe and Kossack. To the program of his new school Worpitzky contributed a research article on function theory in '70. Things continued to go well for him.

In the fall of '72 he received, in addition to his position at the Gymnasium, an appointment to the königliche Kriegsakademie, where he was a colleague of Kummer, Schellbach and later of Lampe. He continued in both positions until his death. In addition, Worpitzky was, from time to time, called in to help at the Ministry of Education. '72 was also the year in which the first volume of his "Elemente der Mathematik für gelehrte Schulen und zum Selbststudium" appeared. Other volumes were published in '74 and '78. A second edition came out in '83.

In '73 Worpitzky was promoted to an "Oberlehrerstelle" and published two articles in the Archiv f. Math. u. Phys. whose editor, after Grunert's death in '72, was now Hoppe. In '75 the title "Professor" was conferred on Worpitzky. His salary at this time was 5400 Mark (= 1800 Thaler). We do not know whether this was for both of his positions or only the one at the Gymnasium. By '76 he had moved up to the "1. Oberlehrerstelle" and received a raise in salary to 6000 Mark. These salaries are comparable to those of university professors at that time. Worpitzky lived in at least eight places during his years in Berlin. That together with the absence of any reference to either a wife or children in the documents available to us make us think that he was a bachelor all his life. Thus he was financially very well off. In '91 his salary was raised to 6300 Mark and he was awarded the "Kronenorden 4. Klasse". In '93 he was further honored by being made a "Rat 4. Klasse" (most Oberlehrer were in the 5. Klasse).

His publications continued. The "Lehrbuch der Differential und Integralrechnung" was published in '80. In '81 he had an article in the Festschrift of his Gymnasium and in '83 the J. f.d.r.u.a. Math (now edited by Kronecker and Weierstrass) printed a thirty page article by him on Bernoulli numbers. Altogether he published two books and seventeen articles, the last in '87.

Most of Worpitzky's publications after '67 were reviewed in the "Jahrbuch über die Fortschritte der Mathematik". Many received lengthy and generally favorable reviews. Thus the six pages devoted to "Diff. u. Int. Rechnung" were quite complimentary. Less well fared his writings on geometry. In particular the third and fourth part of "Elemente" received a devastating critique.

His teaching was praised both at the Gymnasium and at the Kriegsakademie.

Beginning in '92, Lampe reports, "his health deteriorated due to a continuing series of different illnesses so that, because of the slowly progressing destruction of his formerly great powers, he was able to do but very little. In view of this he had asked for retirement effective April 1, 1895, but died a month before this time on March 4, 1895." Lampe then describes Worpitzky's collapse during a lecture at the Kriegsakademie and how he was carried home by a group of orderlies. After several weeks he died of influenza, "which disease had already attacked him repeatedly in the years before and which, each time, weakened him severely."

From reports from his Gymnasium we gather that he was on leave for the SS94 and that in the WS94/5 he also had to stop teaching after a few weeks. Considering that both his father and his sister died of tuberculosis, one wonders whether this was also the cause of his death.

6. Worpitzky's contributions to convergence of continued fractions with complex elements. In '65 Worpitzky published in the "Friedrichs Gymnasium und Realschule Jahresbericht" a 37 page article on "Untersuchungen über die Entwickelung der monodromen und mongenen Functionen durch Kettenbrüche. Erste Folge". Recall that already in the '62 article on "Riccati equations" he had shown great interest in convergence of continued fractions. The main aim of the present paper is to show that the continued fraction

$$(6.1) \qquad \frac{1}{1} + \frac{a_1 z}{1} + \frac{a_2 z}{1} + \cdots,$$

where the elements a_n are complex numbers and z is a complex variable, converges for $|z| < 1/4\,a$, provided $|a_n| \le a$, $n \ge 1$, to the function whose power series expansion formally corresponds to (6.1). Clearly, the result we mentioned in Section 1, that $|a_n| \le 1/4$, $n \ge 1$, insures the convergence of $K(a_n/1)$, is a corollary of Worpitzky's main result.

Instead of (6.1) it is slightly more convenient to use

$$(6.2) \qquad \frac{1}{1} - \frac{a_1 z}{1} - \frac{a_2 z}{1} - \cdots .$$

Worpitzky makes essential use of the "tails" of these continued fractions and of their approximants

$$(6.3) \qquad \phi_n^{(k)}(z) := \frac{1}{1} - \frac{a_k z}{1} - \cdots - \frac{a_{k+n} z}{1} .$$

Each $\phi_n^{(k)}$ is a rational function in z and has a Taylor series expansion

$$\phi_n^{(k)}(z) =: \sum_{v=0}^{\infty} p_n^{(k,n)} z^v, \qquad p_0^{(k,n)} := 1 .$$

It was known before Worpitzky that $p_v^{(k,n)}$ is independent of n for $v = 0, \ldots, n+1$, say

$$p_v^{(k,n)} =: p_v^{(k)}, \qquad v = 0, \ldots, n+1 .$$

In this way to each tail (including the zeroth tail) there corresponds a uniquely determined power series

$$\sum_{v=0}^{\infty} p_v^{(k)} z^v .$$

This series, with $k = 0$, is the one referred to in the statement of the theorem.

Worpitzky's proof begins with an analysis of the coefficients $p_v^{(k)}$ as functions of the a_n

$$p_n^{(k)} = P_n(a_k, \cdots) .$$

He shows that each P_n is a homogeneous polynomial in the variables $a_k, \ldots, a_{k+n-1}$ of degree n with all coefficients positive. By a majorization process (which goes back to Cauchy in other contexts) he then shows that (6.2) converges for $|z| < r$ if the periodic continued fraction

$$(6.4) \qquad \frac{1}{1} - \frac{az}{1} - \frac{az}{1} - \cdots ,$$

where $|a_n| \leq a$, $n \geq 1$, converges for $|z| < r$. The convergence of (6.4) had been established earlier. It is not clear whether Worpitzky was aware of this. He gives a new, ingenious proof for the convergence of (6.4) for $|z| < r = 1/4$ a.

In proving convergence of (6.2) Worpitzky again asserts, incorrectly, that the convergence of $\phi_n^{(0)} - \phi_{n-1}^{(0)}$ to zero is sufficient for the convergence of the sequence $\{\phi_n^{(0)}\}$ of approximants of (6.2). But again his conclusion is correct since

$$|\phi_n^{(0)}(z) - \phi_{n-1}^{(0)}(z)| = O(|4az|^n) .$$

There is much else in the paper which is of interest. Particularly intriguing is the assertion that if $\lim a_n = a$, then the function to which (6.2) converges for $|z| < 1/4 \, |a|$ must have a singular point on the circle $|z| = 1/4 \, |a|$ at least if $a_n > 0$, $n \geq 1$. Unfortunately, his sketchy proof breaks down because he does not account for possible poles of tails of (6.2), which would not lead to singularities of the function itself. Thus, except when $|a_n - a| < r^n$ $\quad 0 < r < 1$, in which case the conjecture is true, the question remains open.

There are more results on the tails of (6.2). These include a "value region" theorem as well as a discussion of the critical tails $S_n^{-1}(\infty)$ which Worpitzky calls "umgekehrte Kettenbrüche". In the limit periodic case, $a_n \to a$ the convergence of $\{S_n^{-1}(\infty)\}$ to $(-1 + \sqrt{1+4az})/2$ is established. These results are obtained for continued fractions of the form

$$(6.5) \qquad \mathop{K}_{n=1}^{\infty} \left(a_n z^{\alpha_n}/1\right) ,$$

where the α_n are positive integers. The article concludes with a discussion of formal expansions of functions into continued fractions of the form (6.5).

It is not clear what Worpitzky meant to take up in later installments (Erste Folge = first install-ment). Except for a section on continued fractions in his textbook "Elemente der Mathematik" and a discussion of two examples in the article on function theory in '70 he did no further work on continued fractions after '65.

7. Why was Worpitzky's result overlooked? Some conjectures.

It has been suggested (by E. T. Bell among others) that mathematical results published in secondary school reports are not likely to attract attention. This proved to be the case for Worpitzky's theorem, but there are numerous rea-sons why a different outcome could have been expected. Mathematical research had appeared in annual publications of secondary schools before '65. Both Kummer and Weierstrass had made use of these outlets for some of their results.

Beginning with the mathematical results published in '68 the Jahrbuch über die Fortschritte der Mathematik set out to review all published mathematical research. This included privately printed pamphlets and certainly publications in annual reports of schools and even textbooks, if they contained some novel approaches. Both of Worpitzky's books and all of his articles after '67 were reviewed in "Fortschritte". Before '68 there was no reviewing journal as thorough as "Fortschritte" but the "Litteraturzeitung" in "Zeitung für Mathematik und Physik" did do a reasonable amount of reviewing. Worpitzky's thesis was reviewed there in vol. 13 (68), p. 6. Unfortunately we have had no access to the volume which might contain a review of his article on convergence of continued fractions.

Similarly, the fact that Worpitzky was only a secondary school teacher should not have stood in the way of his contribution attracting attention. As we saw earlier, many successful mathematicians spent time teaching in secondary schools before being called to university positions. Also the total output of mathematical research as well as the number of active mathematicians was still quite small in '65 so that a significant contribution should not have been overlooked.

A related question, since Worpitzky did not yet have a doctor's degree in '65, is whether the '65 paper or even the '62 paper were initially meant to be dissertations. If he had been thinking along these lines did he submit either of the articles as a thesis? If yes, why was in particular the '65 article not accepted? The two logical choices for submission of a thesis would appear to have been Greifswald or Berlin. Grunert, who was Ordinarius at Greifswald, had been interested in continued fractions and probably had suggested at least the Riccati investigation to Worpitzky. He also had at least one student who wrote a thesis on convergence of continued fractions under him. This was Fr. Arndt in '45. Arndt was at that time a teacher in Stralsund. Later he was a lecturer and eventually an associate professor at UB. Worpitzky may have had a falling out with Grunert. Otherwise it would appear to be strange that Worpitzky published in the Archiv. f. Math. u. Phys. only after Grunert had died and Hoppe had become editor.

Since Arndt was interested in continued fractions and was in addition both a fellow alumnus of Worpitzky and a fellow Pomeranian, it is hard to believe that they did not become acquainted in Berlin. Though Arndt was only an associate professor and thus could not sponsor a doctoral candidate one might have thought that he could have convinced one of the full professors (Ohm, Kummer and Weierstrass at that time) to accept Worpitzky's paper as a dissertation.

One possible stumbling block may have been that theses, at Prussian universities, had to be written in Latin at that time. The first mathematical dissertation written in German was accepted at UB in '72. However, since Worpitzky graduated first in his class from a humanistische Gymnasium, it is likely that he could have met this requirement.

Besides Arndt, Borchardt (lecturer at UB and editor of "Crelle's") and Thomé (lecturer and associate professor at UB) had done work on continued fractions. Thomé, in Berlin since '63 where he received his doctorate in '64, wrote two papers published in "Crelle's" in '66 and '67 on the convergence of continued fraction expansions of certain functions, among others those given by ratios of hypergeometric series. Thomé proved convergence in a slit plane containing a disk with center at the origin. Convergence in this disk follows from Worpitzky's result of '65. It is hard to understand that Thomé did not hear from Worpitzky directly or through some common acquaintance about this fact. That the question of convergence of the continued fraction expansion of the ratio of hypergeometric series, left open by Gauss, was considered to be of some importance is shown both by the fact that Riemann in '64 (in a fragment published post-humously in '76) addressed the problem and that Weierstrass was sufficiently impressed by Thomé to urge him to become a "Privatdozent" at UB in '69 and to support him for the associate professorship which became vacant when Fuchs went to Greifswald in '69.

Politics were not unknown at UB. Weierstrass preferred Thomé over Christoffel for Fuchs' position. This, in retrospect, clearly was the wrong choice. Thomé had an undistinguished career as Ordinarius in Greifswald ('74-'10). Beginning with his first visit to Berlin in '69/70. F. Klein developed an antipathy to Weierstrass which was reciprocated by Weierstrass and his school and caused some friction between Göttingen and Berlin.

Weierstrass also did not have a very high opinion of Schellbach. Among other conflicts between the two there was Weierstrass' opposition to the appointment of Schellbach to the editorship of "Crelle's". To the extent that there was a "Schellbach party", Worpitzky most likely was a member of it. This may help to explain why he was not in the good graces of the powers at UB, if this was indeed so. All this is clearly very conjectural, but Worpitzky's result being so completely overlooked is so hard to understand that one is tempted to engage even in far fetched conjectures.

Worpitzky clearly did not blow his own horn sufficiently. He certainly should have let it be known that he had a good part of Thomé's results and much more, well before the latter published them in "Crelle's". But beyond a modest remark in his article "Beiträge zur Functionen-theorie" published in another school program in '70 he never asserted his priority. In '70 Worpitzky had the following to say:

> In another place is derived from the above that every continued fraction $K(u_n/1)$ converges to the value of the function whose expansion it is, if for increasing n the modulus of u_n remains $< 1/4$. An example is the Gaussian quotient of hypergeometric series.

After '65 Worpitzky's research turned into other paths. There is a section on continued fractions in his "Elemente der Mathematik" (Heft 2, Algebra, Kettenbrüche, etc. 1874). It does not contain a reference to the author's convergence result.

8. Sources and Acknowledgements. For data on Worpitzky's family we are indepted to the Rev. Barsch, Pastor of the "Evangelische Pfarrant Zarnekow (DDR)". Two obituaries for Worpitzky written by E. Lampe (the first published in the "Verhandlungen der physikalischen Gesellschaft zu Berlin" and the second, a slight condensation of the first, appearing in the "Jahresbericht der DMV") provided important material. A list of Worpitzky's publications appended to these obituaries confirmed the information we had already obtained from other sources. Among these the "Jahrbuch über die Fortschritte der Mathematik" was of most importance since it also threw some light on the reaction of the mathematical community to his publications after '67. We are grateful to Professor G. Wechsung of the Friedrich Schiller Universität Jena (DDR) for his help in obtaining general information about the University in '67 as well as all references to Worpitzky in its archieves. He also provided us with a complete and legible copy of Worpitzky's earliest publication (before that we only had an incomplete copy which, in some places, was almost illegible). Dr. Lutz Voelkel and Dr. Peter Schreiber of the Ernst Moritz Arndt Universität, Greifswald also kindly sent us copies of one of Worpitzky's publications, and provided information about the mathematicians in Greifswald between 1818 and 1874.

A list of graduates of the "Gymnasium zu Anclam" of 1855 as well as various documents from the "Pädagogische Zentrum, Berlin", excerpts from the lists of the "Provinzial Schul–Kollegium in Berlin" as well as "Jahresberichte", "Festschriften" and histories of the Friedrichs–Werdersche Gymnasium all helped in reconstructing Worpitzky's vita. In this context the help of Dr. H. G. Schultze Berndt, Verein für die Geschichte Berlins; Frau Ursula Müller, Pädagogisches Zentrum, Berlin, Oberstudiendirektor Dr. Völker Georg–Herwegh–Oberschule, Berlin and Librarian Kjersti Lie, The University of Trondheim, should be acknowledged.

Much of our knowledge about mathematicians in Greifswald and Berlin came from the "Allgemeine Deutsche Biographie" 56 vols., Bayrische Akademie, Dunker und Humblot, München 1875–1912 and the "Neue Deutsche Biographie" Bayrische Akademie, vol. 1 (1952) – vol. 15 (1986), Dunker und Humblot, Berlin. Biographies and other historical articles in the "Jahresberichte der DMV" also proved helpful.

A great deal of our information about UB was obtained from: Kurt R. Biermann, "Die Mathematik und ihre Dozenten an der Berliner Universität, 1810–1920", Akademie Verlag, Berlin 1973. The article: Eberhard Knobloch, "Die Berliner Gewerbeakademie und ihre Mathematiker", in E.B. Christoffel, P.L. Butzer, F. Fehér, eds., pp. 42–51, Birkhäuser Verlag, Basel 1981 provided many facts about that institution.

Finally, one of the present authors (HW) was supported by the Alexander von Humboldt–Stiftung, Bonn–Bad Godesberg, while collecting information on Worpitzky in Berlin.

9. List of Worpitzky's Publications.

Books

1. Elemente der Mathematik für gelehrte Schulen und zum Selbststudium. Berlin, Weidmann. Erstes Heft: Die Arithmetik. 1872. Zweites Heft: Algebra, Combinationslehre nebst Wahrscheinlichkeitsrechnung, Kreisfunctionen nebst Trigonometrie. 1872. Drittes und viertes Heft: Planimetrie. 1874. Fünftes Heft: Stereometrie. 1883.

2. Lehrbuch der Differential-und Integralrechnung. Berlin, Weidmann. 794 S. 1880.

Articles

1. Beitrag zur Integration der Riccati'schen Gleichung. Greifswald, 1862.

2. Ueber die Entwickelung der monodromen und monogenen Functionen durch Kettenbrüche. Erste Folge. Progr. Friedr.-Gymn. Berlin, 1865.

3. Ueber die Endlichkeit von bestimmten Integralen und Reihensummen. Jenenser Inaug.-Dissert. Berlin, 1867.

4. Beiträge zur Functionentheorie. Progr. Friedr. Werder'sches Gymn. Berlin, 1870.

5. Ueber das bestimmte Integral $\int_0^{2\pi} \dfrac{d\phi}{A + B\cos\phi + C\sin\phi}$, in welchem A, B, C beliebige (reelle oder complexe) Constanten sind. Arch. für Mathemat. u. Phys. LV, pp. 59-64. 1873.

6. Ueber die Grundbegriffe der Geometrie. Arch. für Math. u. Phys. LV, pp. 405-421. 1873.

7. Auswerthung des Integrals $\int_0^\infty \dfrac{x^{\alpha-1}}{x+\mu}\, dx$. Zeitschr. für Math. u. Phys. XIX, pp. 90-92. 1874.

8. Ueber die Wurzeln der Gleichungen. Berlin, 1877.

9. On the roots of equations. Analyst V, pp. 51-52. 1878.

10. Ueber die Verallgemeinerung der partiellen Integration. Zeitschr. für Math. u. Phys. XXIII, pp. 407-408. 1878.

11. Zahl, Grösse, Messen. Festschr. Friedr. Werder'sches Gymn. 16 S. 8^0. 1881.

12. Studien über die Bernoulli'schen und Euler'schen Zahlen. Journ. f. d. reine u. angew. Math. XCIV, pp. 203-233. 1883.

13. Ueber die Partialbruchzerlegung der Functionen mit besonderer Anwendung auf die Bernoulli'schen. Zeitschr. für Math. u. Phys. XXIX, pp. 45-54. 1884.

14. Ueber die ganzzahlige Bestimmung von $\sqrt{b^2 - a^3}$ und $\sqrt[3]{b + \sqrt{b^2 - a^3}}$ bei der Auflösung der cubischen Gleichungen. Zeitschr. für math. u. naturw. Unterr. XVI, pp 578-582. 1885.

15. Ueber die pythagoreischen Dreiecke. Zeitschr. für math. u. naturw. Unterr. XVII, p. 256. 1886.

16. Zur Aufstellung Quadratischer Gleichungen $x^2 \pm ax \pm b = 0$, deren Wurzeln bei jeder Combination der Vorzeichen von a und b rational sind. Zeirtschr. für math. u. naturw. Unterr. XVII, pp. 257, 499-501. 1886.

17. Ueber die realen Lösungen der Gleichung $a^\alpha = b^2 + c^2$. Zeitschr. für math. u. naturw. Unterr. XVIII, pp. 168-177. 1887.

LJ:
Institutt for matematiske fag,
Universitetet i Trondheim (NTH),
N7034 Trondheim

WJT:
Department of Mathematics
University of Colorado, Boulder
Boudler, CO 80309-0426

HW:
Institutt for Matematikk og Statistikk
Universitetet i Trondheim (AVH)
N 7055 Dragvoll

Positive T-fraction Expansions for a Family of Special Functions

William B. Jones

Department of Mathematics

University of Colorado at Boulder

Boulder, CO 80309-0426, USA*

Nancy J. Wyshinski

Department of Mathematics

University of Colorado at Denver

Denver, CO 80204-5300, USA*

Abstract: Positive T-fractions are studied for analytic functions of the form

$$F(n, w) := \int_0^w \frac{du}{1 + u^n}, n = 1, 2, 3, \cdots.$$

It is shown that such functions can be expressed as Stieltjes transforms and that the related moments can be computed by means of recurrence relations. The positive T-fraction coefficients are then computed using quotient-difference relations and the moments. Special attention is given to the approximation and computation in the complex plane of the two functions

$$F(1, w) = Log(1 + w) \quad \text{and} \quad F(2, w) = Arctan(w)$$

by approximants $f_m(w)$ of the corresponding positive T-fraction. The rational functions $f_m(w)$ are two-point Padé approximants, and numerical experiments are given using various choices for the two points of interpolation. Contour maps of the number of significant digits $SD(f_m(w))$ in the approximations $f_m(w)$ are used to describe the convergence behavior of the continued fraction at different parts of $\mathcal{C}$ and for different choices of interpolation points.

1. Introduction

In [9], it was shown that functions $G(z)$ defined by *Stieltjes transforms*

$$G(z) := z \int_a^b \frac{d\phi(t)}{z + t}, \qquad 0 \le a < b < +\infty, \tag{1.1}$$

are analytic for $|arg(z)| < \pi$ and can be represented there by positive T-fractions

$$G(z) = \frac{F_1 z}{1 + G_1 z} + \frac{F_2 z}{1 + G_2 z} + \cdots, \quad F_l > 0, G_l > 0, \tag{1.2}$$

whose approximants, $f_m(z)$, form the main diagonal in the two-point Padé table of $G(z)$. Here $\phi(t) \in \Phi(a, b)$, where $\Phi(a, b)$ denotes the set of all bounded, nondecreasing functions $\phi(t)$, which have infinitely many points of increase on (a, b) and for which the moments

$$c_k := \int_a^b (-t)^k d\phi(t) \tag{1.3}$$

*Research supported in part by the U.S. National Science Foundation under Grant #DMS-8700498.

exist for all integer values of $k = 0, \pm1, \pm2, \cdots$. The positive T-fraction (1.2) converges uniformly on every compact subset of $D(a,b) := \{z : z \notin [-b, -a]\}$ to $G(z)$. The series

$$L_0 = L_0(z) = \sum_{j=1}^{\infty} -c_{-j}z^j, \quad \text{and} \quad L_\infty = L_\infty(z) = \sum_{j=0}^{\infty} c_j z^{-j} \tag{1.4}$$

are asymptotic power series expansions of $G(z)$ at $z = 0$ and $z = \infty$, respectively, with respect to the sectors $R_\alpha = \{z : |Arg\ z| < \alpha, 0 < \alpha < \pi\}$. That is, for each $m = 1, 2, 3, \cdots$,

$$z^{-m}\Big[G(z) - L_{0,m}(z)\Big] = O(z), \quad z \to 0, z \in R_\alpha, \tag{1.5b}$$

and

$$z^m\Big[G(z) - L_{\infty,m}(z)\Big] = O\Big(\frac{1}{z}\Big), \quad z \to \infty, z \in R_\alpha, \tag{1.5b}$$

where

$$L_{0,m} := \sum_{j=1}^{m} -c_{-j}z^j, \quad \text{and} \quad L_{\infty,m} := \sum_{j=0}^{m} c_j z^{-j}. \tag{1.5c}$$

The (m,m) two-point Padé approximant of order $(m+1, m)$ of (L_0, L_∞) is the m-th approximant $f_m(z)$ of the positive T-fraction (1.2), since

$$\Lambda_0(G(z) - f_m(z)) = O_+(z^{m+1}), \quad \text{and} \quad \Lambda_\infty(G(z) - f_m(z)) = O_-(z^{-m}), \tag{1.6}$$

where the symbol $\Lambda_0(f)$ $(\Lambda_\infty(f))$ denotes the Taylor (Laurent) series expansion of f about $z = 0$ $(z = \infty)$. We use the symbol $O_+(z^r)$ to denote a formal (or convergent) power series in increasing powers of z starting with a power of z not less than r. Similarly $O_-(z^{-r})$ denotes a formal (or convergent) power series in decreasing powers of z starting with a power of z less than or equal to $-r$. It can be seen from (1.5) and (1.6) that

$$z^{-m}\Big[G(z) - f_m(z)\Big] = O(z) \quad z \to 0, z \in R_\alpha, \tag{1.7a}$$

and

$$z^{m-1}\Big[G(z) - f_m(z)\Big] = O\Big(\frac{1}{z}\Big) \quad z \to \infty, z \in R_\alpha. \tag{1.7b}$$

The coefficients F_l, G_l are positive numbers that can be computed by the (quotient-difference) FG-relations of McCabe and Murphy given below.

Let $\phi(t) \in \Phi(a,b)$ where $0 \le a < b \le +\infty$. Let L_0 and L_∞ be defined as in (1.4). To compute $F_l, G_l, l = 1, 2, \cdots, p$, set

$$F_1^{(m)} = 0, \quad G_1^{(m)} = -\frac{c_{-m-1}}{c_{-m}}, \quad m = -p, -p+1, \cdots, p-1. \tag{1.8a}$$

Then, for $l = 1, 2, \cdots, p-1$, compute

$$F_{l+1}^{(m)} = F_l^{(m+1)} + G_l^{(m+1)} - G_l^{(m)}, \quad m = l-p-1, l-p, \cdots, p-l-1, \tag{1.8b}$$

$$G_{l+1}^{(m)} = \frac{F_{l+1}^{(m)}}{F_{l+1}^{(m-1)}} G_l^{(m-1)}, \quad m = l-p, l-p+1, \cdots, p-l-1. \tag{1.8c}$$

Finally, set

$$F_1 = -c_{-1}, \qquad F_l = F_l^{(0)}, \; l = 2, 3, \cdots, p, \tag{1.9a}$$

$$G_l = G_l^{(0)}, \; l = 1, 2, \cdots, p. \tag{1.9b}$$

Equations (1.8) are called the *FG-relations* [10].

Truncation error bounds for S-fractions due to Henrici and Pfluger [6] are given by:

Theorem 1.1 *If the S-fraction*

$$\mathbf{K}_{n=1}^{\infty}\left(\frac{a_n z}{1}\right), \; a_n > 0, \; n \geq 1, \tag{1.10}$$

with m-th approximant $g_m(z)$ *converges to a finite value* $G(z)$, *then for* $m = 2, 3, 4, \cdots$,

$$|G(z) - g_m(z)| \leq \begin{cases} |(g_m(z) - g_{m-1}(z))|, & if \; 0 \leq |arg(z)| \leq \dfrac{\pi}{2}; \\ \csc |arg(z)||(g_m(z) - g_{m-1}(z))|, & if \; \dfrac{\pi}{2} < |arg(z)| < \pi. \end{cases} \tag{1.11}$$

An analogous theorem for truncation error bounds for positive T-fractions due to Gragg [4], Jefferson [7] and Jones [8] is given by:

Theorem 1.2 *If the positive T-fraction (1.2) with m-th approximant* $g_m(z)$ *converges to a finite value* $G(z)$, *then for* $m = 2, 3, 4, \cdots$,

$$|G(z) - g_m(z)| \leq \begin{cases} |(g_m(z) - g_{m-1}(z))|, & if \; 0 \leq |arg(z)| \leq \dfrac{\pi}{2}; \\ \csc |arg(z)||(g_m(z) - g_{m-1}(z))|, & if \; \dfrac{\pi}{2} < |arg(z)| < \pi. \end{cases}$$

In this paper, we consider the family of functions

$$F(n, w) := \int_0^w \frac{du}{1 + u^n}, \; n = 1, 2, 3, \cdots, \tag{1.12a}$$

$$w \in D_n := \{w \in \mathcal{C} : w \neq \tau e^{i\frac{\pi(2k+1)}{n}}, \tau \geq 1, k = 0, 1, 2, \cdots, n-1\}, \tag{1.12b}$$

where the path of integration in (1.12a) is the straight line segment from 0 to w. In Section 2, $F(n, w)$ is expressed (Theorem 2.1) in terms of Stieltjes transforms (1.1) so that positive T-fractions converging to (1.2) can be obtained. Recursion formulas used to compute the moments (1.3) are developed in Theorems 2.2 and 2.3. With the moments c_k we compute the T-fraction coefficients F_l and G_l by means of (1.8). Continued fraction representations of $F(n, w)$ are described by Theorem 2.3. Section 3 is devoted to the study of positive T-fraction representations of the two special functions

$$F(1, w) := \int_0^w \frac{du}{1 + u} = Log(1 + w) \tag{1.13}$$

and

$$F(2,w) := \int_0^w \frac{du}{1+u^2} = Arctan(w) \tag{1.14}$$

where in each case we consider the principal branch. For each function we give numerical values of the moments c_k and coefficients F_l and G_l. These coefficients are used to compute level curves of the number of significant digits $SD(f_m(\alpha,\beta,n,w))$ obtained in the approximation of $F(n,w)$ by an m-th approximant $f_m(\alpha,\beta,n,w)$ of the positive T-fraction. Such maps provide information about the convergence behavior of the continued fractions in various regions of the complex plane. As a means for obtaining an independent check for computing the number of significant digits $SD(f_m(\alpha,\beta,n,w))$, we make use of the well known S-fraction representations

$$F(n,w) = \int_0^w \frac{du}{1+u^n}$$

$$= \frac{w}{1} + \frac{1^2 w^n}{n+1} + \frac{(1 \cdot n)^2 w^n}{2n+1} + \frac{(n+1)^2 w^n}{3n+1} + \frac{(2n)^2 w^n}{4n+1} + \frac{(2n+1)^2 w^n}{5n+1} + \cdots, \tag{1.15}$$

where $n = 1,2,3,\cdots$, and $w \in D_n$. Also used for that purpose are the truncation error bounds given by Theorems 1.1 and 1.2. Finally in Section 4 a relationship is given between the moments c_k of the Stietjes transform and the incomplete beta functions.

2. Continued Fraction Representation of F(n,w)

In Theorem 2.1 $F(n,w)$ is expressed in terms of Stieltjes transforms. Theorems 2.2a and 2.2b both give, for certain functions $\phi(t)$, recurrence formulae for the moments c_k. The continued fraction coefficients F_l and G_l are then computed by the FG-relations (1.8) and (1.9).

Theorem 2.1 *For each $n = 1,2,3,\cdots$, the function $F(n,w)$ defined by (1.12) can be written in the form*

$$F(n,w) = \begin{cases} \dfrac{w\alpha^{1-\frac{1}{n}}}{n(\alpha - \beta w^n)} G\left(n, \dfrac{\alpha}{w^n} - \beta\right), & \text{if } \alpha,\beta > 0, \\[3mm] \dfrac{w\alpha^{1-\frac{1}{n}}}{n(\alpha - \beta w^n)} G\left(n, \beta - \dfrac{\alpha}{w^n}\right), & \text{if } 0 < \alpha \leq |\beta|, \beta < 0, \end{cases}$$

where $w \in D_n$ and

$$G(n,z) := \begin{cases} z \displaystyle\int_\beta^{\alpha+\beta} \dfrac{(t-\beta)^{\frac{1}{n}-1} dt}{z+t}, & \text{if } \alpha,\beta > 0, \\[3mm] z \displaystyle\int_{-\alpha-\beta}^{-\beta} \dfrac{(-\beta-t)^{\frac{1}{n}-1} dt}{z+t}, & \text{if } 0 < \alpha \leq |\beta|, \beta < 0. \end{cases}$$

Proof: The theorem follows from (1.12) by using the successive transformations $u = \frac{wv}{\alpha^{\frac{1}{n}}}$, $v = \tau^{\frac{1}{n}}$, and $t = \tau + \beta$.

Theorem 2.2a *Let α, $\beta > 0$ and a positive integer n be given. Let $\phi_n(t) := n(t-\beta)^{\frac{1}{n}}$, $\beta < t < \alpha + \beta$,. Then the moments c_k defined by*

$$c_k := \int_\beta^{\alpha+\beta} (-t)^k d\phi_n(t) = (-1)^k \int_\beta^{\alpha+\beta} t^k (t-\beta)^{\frac{1}{n}-1} dt \tag{2.1a}$$

exist for all integers $k = 0, \pm 1, \pm 2, \cdots$ and can be calculated using the following recursive formulae:

$$c_k = \frac{1}{1 + kn}\left\{(-1)^k n(\alpha + \beta)^k \alpha^{\frac{1}{n}} - \beta kn c_{k-1}\right\}, \quad k = 1, 2, 3, \cdots, \tag{2.1b}$$

and

$$c_{-k-1} = \frac{-1}{\beta kn}\left\{(-1)^k n(\alpha + \beta)^{-k}\alpha^{\frac{1}{n}} + (kn - 1)c_{-k}\right\}, \quad k = 1, 2, 3, \cdots, \tag{2.1c}$$

where

$$c_0 = n\alpha^{\frac{1}{n}}, \tag{2.1d}$$

$$c_{-1} = \begin{cases} \left\{2\beta^{\frac{1}{n}-1}\displaystyle\sum_{j=0}^{\frac{n}{2}-1} P_j \cos\left(\frac{2j+1}{n}\right)\pi - 2\beta^{\frac{1}{n}-1}\displaystyle\sum_{j=0}^{\frac{n}{2}-1} Q_j \sin\left(\frac{2j+1}{n}\right)\pi\right\}\Big]_0^{(\frac{\alpha}{\beta})^{\frac{1}{n}}}, \\ \qquad \text{if } n \text{ is even;} \\ \left\{-\beta^{\frac{1}{n}-1}\ell n(1 + x) + 2\beta^{\frac{1}{n}-1}\displaystyle\sum_{j=0}^{\frac{n-3}{2}} P_j \cos\left(\frac{2j+1}{n}\right)\pi \right. \\ \left. \qquad -2\beta^{\frac{1}{n}-1}\displaystyle\sum_{j=0}^{\frac{n-3}{2}} Q_j \sin\left(\frac{2j+1}{n}\right)\pi\right\}\Big]_0^{(\frac{\alpha}{\beta})^{\frac{1}{n}}}, \\ \qquad \text{if } n \text{ is odd.} \end{cases} \tag{2.1e}$$

$$P_j = \frac{1}{2}\ell n(x^2 - 2x \cos(\frac{2j+1}{n})\pi + 1),$$

and

$$Q_j = \tan^{-1}\left(\frac{x - \cos(\frac{2j+1}{n})\pi}{\sin(\frac{2j+1}{n})\pi}\right).$$

Proof: Formula (2.1c) for c_0 follows directly from (2.1a). Using (2.1a) and the substitution $t = \beta(x^n + 1)$ yields

$$c_{-1} = -n\beta^{\frac{1}{n}-1}\int_0^{(\frac{\alpha}{\beta})^{\frac{1}{n}}} \frac{dx}{1 + x^n}.$$

This integral can be evaluated by considering the partial fraction decomposition of $\frac{1}{1+x^n}$ to obtain (2.1d). The recurrence formula (2.1b) follows from integration by parts with

$$u = t^k, \ k \neq 0 \text{ and } v = n(t - \beta)^{\frac{1}{n}}.$$

The recurrence formula (2.1c) is an immediate consequence of (2.1b).

Remark. The recurrence formula (2.1b) is used to compute c_k for positive $k \geq 1$; (2.1c) is used to compute c_k for negative $k \leq -2$.

Theorem 2.2b *Let $\beta < 0, \ 0 < \alpha \leq |\beta|$ and a positive integer n be given. Let $\phi_n(t) := n(t - \beta)^{\frac{1}{n}}, -\alpha - \beta < t < -\beta$. Then the moments*

$$c_k := \int_\beta^{\alpha+\beta} (-t)^k d\phi_n(t) = \int_{-\alpha-\beta}^{-\beta} t^k(-\beta - t)^{\frac{1}{n}-1} dt \tag{2.2a}$$

exist for all integers $k = 0, \pm 1, \pm 2, \cdots$, and can be calculated using the following recursive formulae:

$$c_k = \frac{1}{1+kn}\left\{ n(-\alpha - \beta)^k \alpha^{\frac{1}{n}} - \beta kn c_{k-1} \right\}, \quad k = 1, 2, 3, \cdots, \tag{2.2b}$$

and

$$c_{-k-1} = \frac{1}{\beta kn}\left\{ -n(-\alpha - \beta)^{-k} \alpha^{\frac{1}{n}} + (1 - kn)c_{-k} \right\}, \quad k = 1, 2, 3, \cdots, \tag{2.2c}$$

where

$$c_0 = n\alpha^{\frac{1}{n}}, \tag{2.2d}$$

$$c_{-1} = \begin{cases} \left\{ (-\beta)^{\frac{1}{n}-1} \ell n(\frac{1+x}{1-x}) - 2(-\beta)^{\frac{1}{n}-1} \sum_{j=1}^{\frac{n}{2}-1} P_j \cos\left(\frac{2j}{n}\right)\pi + \right. \\ \left. 2(-\beta)^{\frac{1}{n}-1} \sum_{j=1}^{\frac{n}{2}-1} Q_j \sin\left(\frac{2j}{n}\right)\pi \right\}\Big]_0^{(\frac{-\alpha}{-\beta})^{\frac{1}{n}}}, \\ \textit{if } n \textit{ is even;} \\[2mm] \left\{ -(-\beta)^{\frac{1}{n}-1} \ell n(1 - x) + 2(-\beta)^{\frac{1}{n}-1} \sum_{j=0}^{\frac{n-3}{2}} R_j \cos\left(\frac{2j+1}{n}\right)\pi + \right. \\ \left. 2(-\beta)^{\frac{1}{n}-1} \sum_{j=0}^{\frac{n-3}{2}} S_j \sin\left(\frac{2j+1}{n}\right)\pi \right\}\Big]_0^{(\frac{-\alpha}{-\beta})^{\frac{1}{n}}}, \\ \textit{if } n \textit{ is odd.} \end{cases} \tag{2.2e}$$

$$P_j = \frac{1}{2}\ell n(x^2 - 2x\cos(\frac{2j}{n})\pi + 1),$$

$$Q_j = \tan^{-1}\left(\frac{x - \cos(\frac{2j}{n})\pi}{\sin(\frac{2j}{n})\pi}\right),$$

$$R_j = \frac{1}{2}\ell n(x^2 + 2x\cos(\frac{2j+1}{n})\pi + 1),$$

and

$$S_j = \tan^{-1}\left(\frac{x + \cos(\frac{2j+1}{n})\pi}{\sin(\frac{2j+1}{n})\pi}\right).$$

Proof: As in the preceding proof, c_0 follows directly from (2.2a). Using (2.2a) and the substitution $t = \beta(x^n - 1)$ yields

$$c_{-1} = n(-\beta)^{\frac{1}{n}-1} \int_0^{(\frac{-\alpha}{-\beta})^{\frac{1}{n}}} \frac{dx}{1 - x^n}.$$

By considering the partial fraction decomposition of $\frac{1}{1-x^n}$, the above integral can be evaluated and then (2.2d) is obtained. The recurrence formula follows using integration by parts with

$$u = t^k, \quad k \neq 0, \text{ and } v = -n(-\beta - t)^{\frac{1}{n}}.$$

Remark. The recurrence formula (2.2b) is used to compute c_k for positive $k \geq 1$; (2.2c) is used to compute c_k for negative $k \leq -2$.

Theorem 2.3 *The function $F(n,w)$, where α and β are as in Theorems 2.2a and 2.2b, is represented in*

$$\begin{cases} D(\beta, \alpha + \beta), & \text{if } \alpha, \beta > 0, \\ D(-\alpha - \beta, -\beta), & \text{if } 0 < \alpha \le |\beta|, \beta < 0, \end{cases}$$

by a continued fraction

$$f(\alpha, \beta, n, w) := \begin{cases} \dfrac{w\alpha^{1-\frac{1}{n}}}{n(\alpha - \beta w^n)} \left[\dfrac{F_1 \cdot (\frac{\alpha}{w^n} - \beta)}{1 + G_1 \cdot (\frac{\alpha}{w^n} - \beta)} + \dfrac{F_2 \cdot (\frac{\alpha}{w^n} - \beta)}{1 + G_2 \cdot (\frac{\alpha}{w^n} - \beta)} + \cdots \right], \\ \qquad \text{if } \alpha, \beta > 0, \\ \dfrac{w\alpha^{1-\frac{1}{n}}}{n(\alpha - \beta w^n)} \left[\dfrac{F_1 \cdot (\beta - \frac{\alpha}{w^n})}{1 + G_1 \cdot (\beta - \frac{\alpha}{w^n})} + \dfrac{F_2 \cdot (\beta - \frac{\alpha}{w^n})}{1 + G_2 \cdot (\beta - \frac{\alpha}{w^n})} + \cdots \right], \\ \qquad \text{if } 0 < \alpha \le |\beta|, \ \beta < 0. \end{cases}$$

The positive coefficients F_l and G_l are defined by the FG-relations (1.8). The continued fraction converges uniformly on every compact subset of

$$\begin{cases} D(\beta, \alpha + \beta), & \text{if } \alpha, \beta > 0, \\ D(-\alpha - \beta, -\beta), & \text{if } 0 < \alpha \le |\beta|, \beta < 0, \end{cases}$$

The expression in brackets is a positive T-fraction in

$$\begin{cases} z = \frac{\alpha}{w^n} - \beta, & \text{if } \alpha, \beta > 0, \\ z = \beta - \frac{\alpha}{w^n}, & \text{if } 0 < \alpha \le |\beta|, \ \beta < 0. \end{cases}$$

Proof: This result follows directly from [9, Theorem 2.1].

Recall that the approximants $f_m(z)$ of a positive T-fraction have the asymptotic properties at $z = 0$ and $z = \infty$ given by (1.7). It follows that with $\alpha, \beta > 0$ and with $z = \frac{\alpha}{w^n} - \beta$, $f_m(\frac{\alpha}{w^n} - \beta)$ have asymptotic properties of $G(n, \frac{\alpha}{w^n} - \beta)$ about $w = 0$ and $w = (\frac{\alpha}{\beta})^{\frac{1}{n}}$, in the sense that

$$\left(\frac{\alpha}{w^n} - \beta \right)^{-m} \left[G(n, \frac{\alpha}{w^n} - \beta) - f_m(\frac{\alpha}{w^n} - \beta) \right] = O\left(\frac{\alpha}{w^n} - \beta \right), \quad w \to \left(\frac{\alpha}{\beta} \right)^{\frac{1}{n}},$$

and

$$\left(\frac{\alpha}{w^n} - \beta \right)^{m} \left[G(n, \frac{\alpha}{w^n} - \beta) - f_m(\frac{\alpha}{w^n} - \beta) \right] = O\left(\frac{1}{\frac{\alpha}{w^n} - \beta} \right), \quad w \to 0,$$

for $z \in D(\beta, \alpha + \beta)$. From this it is clear that the approximation of $F(n, w)$ by $f_m(z)$ will be good near the interpolation points $w = 0$ and $w = (\frac{\alpha}{\beta})^{\frac{1}{n}}$. The other case when $0 < \alpha \le |\beta|, \beta < 0$ is handled in a similar fashion.

3. Examples

In this section, we consider in greater detail two members of the family of integrals discussed in Section 2.

Example 1. $n = 1$. The logarithm function

$$F(1, w) = Log(1 + w) = \int_0^w \frac{du}{1 + u}$$

can be written in the form

$$F(1, w) = \begin{cases} \dfrac{w}{(\alpha - \beta w)} G\left(1, \dfrac{\alpha}{w} - \beta\right), & \text{if } \alpha, \beta > 0, \\[2mm] \dfrac{w}{(\alpha - \beta w)} G\left(1, \beta - \dfrac{\alpha}{w}\right), & \text{if } 0 < \alpha \le |\beta|, \beta < 0, \end{cases}$$

where

$$G(1, z) := \begin{cases} z \displaystyle\int_\beta^{\alpha + \beta} \dfrac{dt}{z + t}, & \text{if } \alpha, \beta > 0, \\[4mm] z \displaystyle\int_{-\alpha - \beta}^{-\beta} \dfrac{dt}{z + t}, & \text{if } 0 < \alpha \le |\beta|, \beta < 0, \end{cases}$$

for all $w \in D_1$. D_1 is the complex plane cut along the negative real axis from -1 to $-\infty$. Thus, $F(1, w)$ can be represented by a positive T-fraction

$$f(\alpha, \beta, 1, w) := \begin{cases} \dfrac{w}{(\alpha - \beta w)} \left[\dfrac{F_1 \cdot (\frac{\alpha}{w} - \beta)}{1 + G_1 \cdot (\frac{\alpha}{w} - \beta)} + \dfrac{F_2 \cdot (\frac{\alpha}{w} - \beta)}{1 + G_2 \cdot (\frac{\alpha}{w} - \beta)} + \cdots \right], & \text{if } \alpha, \beta > 0, \\[4mm] \dfrac{w}{(\alpha - \beta w)} \left[\dfrac{F_1 \cdot (\beta - \frac{\alpha}{w})}{1 + G_1 \cdot (\beta - \frac{\alpha}{w})} + \dfrac{F_2 \cdot (\beta - \frac{\alpha}{w})}{1 + G_2 \cdot (\beta - \frac{\alpha}{w})} + \cdots \right], & \text{if } 0 < \alpha \le |\beta|, \\[2mm] & \beta < 0. \end{cases}$$

(3.1)

To describe the convergence behavior of $f(\alpha, \beta, 1, w)$ we consider contour maps of the number of significant digits, $SD(f_m(\alpha, \beta, 1, w))$, in the approximations of $Log(1 + w)$ by the mth approximant $f_m(\alpha, \beta, 1, w)$ of (3.1) with appropriate α and β. To compute the number of significant digits, we use the approximation formula

$$SD(\widetilde{W}) \simeq \widetilde{SD}(\widetilde{W}) := -Log_{10} \left| \frac{\widetilde{W} - W}{W} \right| - .301$$

(3.2)

where $\widetilde{W}$ is an approximation to W. $W = Log(1 + w)$ is calculated to machine precision using the principal value of $Log(1 + w) = \ell n|1 + w| + iArg(1 + w)$. All computations were made on the University of Colorado's CDC Cyber using double precision, so that the maximum number of significant digits is about 28. All plots were produced using the CONREC plotting package from the National Center for Atmospheric Research, using a resolution of 100. Since the function $G(1, z)$ has a positive T-fraction which corresponds at $z = 0$ and $z = \infty$, the resulting approximants of $f(\alpha, \beta, 1, w)$ will give best numerical results about $w = 0$ and $w = \frac{\alpha}{\beta}$.

Level curves of $\widetilde{SD}(f_{10}(\alpha, \beta, 1, w))$ are shown in Figure 1 for 4 different choices of the parameters α and β and regions of $\mathcal{C}$. In every case $\widetilde{SD}(f_{10}(\alpha, \beta, 1, w))$ has least values near the logarithmic branch point $w = -1$ and the branch cut $-\infty < w < -1$. The values of $\widetilde{SD}(f_{10}(\alpha, \beta, 1, w))$ are seen to increase as w approaches the two interpolation points of the two-point Padé approximant. In Figure 1a the interpolation points are at $w = 0$ and $w = 1$ and the level curves indicate a "surface" rising to sharp peaks above these two points. In Figure 1b we

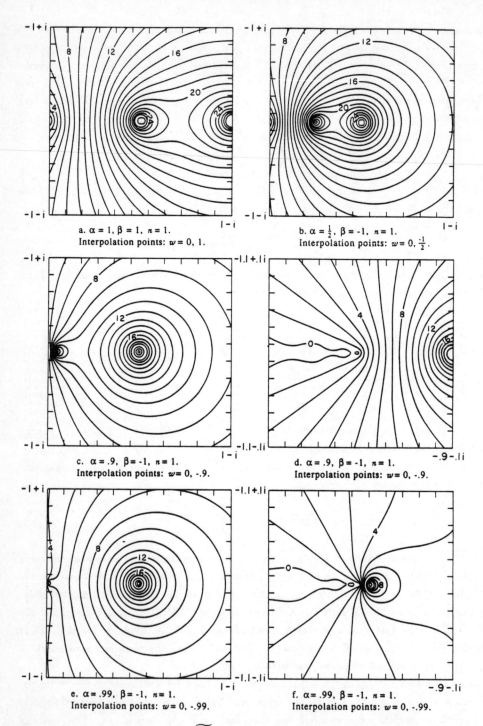

Figure 1. Level curves of the number $\widetilde{SD}(f_{10}(\alpha,\beta,1,w))$ of significant digits in the approximation of $F(1,w) = Log(1 + w)$ by the positive T-fraction approximant $f_{10}(\alpha,\beta,1,w)$.

have kept one interpolation point at $w = 0$ and moved the other to $w = -\frac{1}{2}$. A corresponding change takes place on the surface surrounding the two peaks at $w = 0, -\frac{1}{2}$. In Figures 1c and 1.d the interpolation points are $w = 0$ and $w = -.9$ and the regions shown are different. The region in Figure 1c has the branch point $w = -1$ on the left boundary. In Figure 1d the branch point lies near the center of the region. The scale used in Figure 1d has been increased to show in more detail what occurs near the branch point and cut. Figures 1e and 1f have the same relationship to each other as Figures 1c and 1d; however, in this case one of the interpolation points $w = -.99$ lies closer to the branch point.

Table 1 contains the numerical values of the moments c_k, $k = -10, -9, \cdots, 9, 10$, and the positive continued fraction coefficient F_l and G_l, $l = 1, 2, \cdots, 10$ defined by (1.8) and (1.9) for values of α and β corresponding to those used in Figure 1.

Table 1

a. $\alpha = 1, \beta = 1.$ b. $\alpha = \frac{1}{2}, \beta = -1.$

K	MOMENTS	K	MOMENTS
-10	.11089409722222222222222222D+00	-10	.56777777777777777777777778D+02
-9	-.12451171875000000000000000D+00	-9	.31875000000000000000000000D+02
-8	.14174107142857142857142857D+00	-8	.18142857142857142857142857D+02
-7	-.16406250000000000000000000D+00	-7	.10500000000000000000000000D+02
-6	.19375000000000000000000000D+00	-6	.62000000000000000000000000D+01
-5	-.23437500000000000000000000D+00	-5	.37500000000000000000000000D+01
-4	.29166666666666666666666667D+00	-4	.23333333333333333333333333D+01
-3	-.37500000000000000000000000D+00	-3	.15000000000000000000000000D+01
-2	.50000000000000000000000000D+00	-2	.10000000000000000000000000D+01
-1	-.69314718055994530941723212D+00	-1	.69314718055994530941723212D+00
0	.10000000000000000000000000D+01	0	.50000000000000000000000000D+00
1	-.15000000000000000000000000D+01	1	.37500000000000000000000000D+00
2	.23333333333333333333333333D+01	2	.29166666666666666666666667D+00
3	-.37500000000000000000000000D+01	3	.23437500000000000000000000D+00
4	.62000000000000000000000000D+01	4	.19375000000000000000000000D+00
5	-.10500000000000000000000000D+02	5	.16406250000000000000000000D+00
6	.18142857142857142857142857D+02	6	.14174107142857142857142857D+00
7	-.31875000000000000000000000D+02	7	.12451171875000000000000000D+00
8	.56777777777777777777777778D+02	8	.11089409722222222222222222D+00
9	-.10230000000000000000000000D+03	9	.99902343750000000000000000D-01
10	.18609090909090909090909091D+03	10	.90864701704545454545454545D-01

	F_l		F_l
L		L	
1	.69314718055994530941723212D+00	1	.69314718055994530941723212D+00
2	.28200339884536394262730219D-01	2	.56400679769072788525460438D-01
3	.22948410519223648605084174D-01	3	.45896821038447297210168347D-01
4	.22069485016134388942270690D-01	4	.44138970032268777884541380D-01
5	.21790531538164492679248296D-01	5	.43581063076328985358494727D-01
6	.21664865594656897308777378D-01	6	.43329731189313794618111752D-01
7	.21597473353819903540712730D-01	7	.43194946707639806970374793D-01
8	.21557130848306832421576070D-01	8	.43114261696613683409628575D-01
9	.21531065646876193658203785D-01	9	.43062131293749576469718687D-01
10	.21513250158023800150730490D-01	10	.43026500316446687319863498D-01

	G_l		G_l
L		L	
1	.69314718055994530941723212D+00	1	.13862943611199806188344642D+01
2	.70996456735920276079491987D+00	2	.14199291347184055215898397D+01
3	.70749391006621008154943459D+00	3	.14149878201324201630988692D+01
4	.70723818301906804948305001D+00	4	.14144763660381360989661019D+01
5	.70716666987120567510146519D+00	5	.14143333397424113502024358D+01
6	.70713906614369079041895657D+00	6	.14142781322873815809302956D+01
7	.70712616369797563710530253D+00	7	.14142523273959512593734501D+01
8	.70711932722227484929025776D+00	8	.14142386544445518881723121D+01
9	.70711536725915470291440739D+00	9	.14142307345180033597712137D+01
10	.70711291500093504512049929D+00	10	.14142258300430441796432554D+01

For Table 1a, $\alpha = 1$ and $\beta = 1$; for Table 1b, $\alpha = \frac{1}{2}$ and $\beta = -1$. Since the same values of α and β are used in Figures 1c and 1d, we denote the corresponding table as Table 1c,d. Therefore for Table 1c,d, $\alpha = .9$ and $\beta = -1$. Similarly, since $\alpha = .99$ and $\beta = -1$ for Figures 1e and 1f, Table 1e,f contains the moments and positive continued fraction coefficients for $\alpha = .99$ and $\beta = -1$.

Table 1

c,d. $\alpha = .9, \beta = -1$.   e,f. $\alpha = .99, \beta = -1$.

K	MOMENTS	K	MOMENTS
-10	.111111111100000000000000000D+09	-10	.111111111111111111100000000D+18
-9	.124999998750000000000000000D+08	-9	.124999999999999998750000000D+16
-8	.142857128571428571428571430D+07	-8	.142857142857142857142857D+14
-7	.166666500000000000000000000D+06	-7	.166666666666650000000000000D+12
-6	.199998000000000000000000000D+05	-6	.199999999980000000000000000D+10
-5	.249975000000000000000000000D+04	-5	.249999997500000000000000000D+08
-4	.333000000000000000000000000D+03	-4	.333333000000000000000000000D+06
-3	.495000000000000000000000000D+02	-3	.499950000000000000000000000D+04
-2	.900000000000000000000000000D+01	-2	.990000000000000000000000000D+02
-1	.230258509299404568401799150D+01	-1	.460517018598809136803598290D+01
0	.900000000000000000000000000D+00	0	.990000000000000000000000000D+00
1	.495000000000000000000000000D+00	1	.499950000000000000000000000D+00
2	.333000000000000000000000000D+00	2	.333333000000000000000000000D+00
3	.249975000000000000000000000D+00	3	.249999997500000000000000000D+00
4	.199998000000000000000000000D+00	4	.199999999980000000000000000D+00
5	.166666500000000000000000000D+00	5	.166666666666650000000000000D+00
6	.142857128571428571428571430D+00	6	.142857142857142857142857D+00
7	.124999998750000000000000000D+00	7	.124999999999999998750000000D+00
8	.111111111100000000000000000D+00	8	.111111111111111111100000000D+00
9	.999999999900000000000000000D-01	9	.999999999999999999999000000D-01
10	.909090909081818181818182D-01	10	.909090909090909090909000000D-01

F_l F_l

L		L	
1	.230258509299404568401799150D+01	1	.460517018598809136803598290D+01
2	.135022245602477124439572530D+01	2	.168458897976573378242500910D+02
3	.129027295301234925329006440D+01	3	.245241732712159911565936130D+02
4	.120602717486370982005877270D+01	4	.206684770669554413851634430D+02
5	.118916182205280425443224790D+01	5	.207138916471110777194586200D+02
6	.118145937554229488742900784D+01	6	.204711211237358653682285950D+02
7	.117747304151842362493894230D+01	7	.204140016749720655548889663D+02
8	.117512204379070746612179822D+01	8	.203641822921630005527061010D+02
9	.117361953981362154768965950D+01	9	.203370709825523260275252710D+02
10	.117260048332323803906389500D+01	10	.203177693311629093105353560D+02

G_l G_l

L		L	
1	.255842788110449520446443490D+01	1	.465168705655362764448079080D+01
2	.331639713199186893413722130D+01	2	.124867432208087162741415270D+02
3	.317246145181447151553554120D+01	3	.974555639902295344875930150D+01
4	.316774240690452748101896160D+01	4	.101210268235475519900087110D+02
5	.316465766517807947964359120D+01	5	.999878296149235534514878900D+01
6	.316358323054294592623553885D+01	6	.100147447476264860632682380D+02
7	.316306399396110269815280570D+01	7	.100048957691699048337007940D+02
8	.316278808648660723319779530D+01	8	.100042503205361460138828740D+02
9	.316262758853710555291521430D+01	9	.100026441734583548095285770D+02
10	.316252795279601593868212640D+01	10	.100019674504258249682157120D+02

Example 2. $n = 2$. The Arctangent function

$$F(2, w) := \int_0^w \frac{du}{1 + u^2} = Arctan(w)$$

is analytic for all $w \in D_2$. We note that D_2 is the complex plane $\mathbb{C}$ cut along the imaginary axis from $+i$ to $+i\infty$ and from $-i$ to $-i\infty$. $F(2, w)$ can be written in the form

$$F(2, w) = \begin{cases} \dfrac{w\alpha^{\frac{1}{2}}}{2(\alpha - \beta w^2)} G\left(2, \dfrac{\alpha}{w^2} - \beta\right), & \text{if } \alpha, \beta > 0, \\[3mm] \dfrac{w\alpha^{\frac{1}{2}}}{2(\alpha - \beta w^2)} G\left(2, \beta - \dfrac{\alpha}{w^2}\right), & \text{if } 0 < \alpha \le |\beta|, \beta < 0, \end{cases}$$

where

$$G(2, z) := \begin{cases} z \displaystyle\int_\beta^{\alpha+\beta} \dfrac{(t - \beta)^{\frac{-1}{2}} dt}{z + t}, & \text{if } \alpha, \beta > 0, \\[4mm] z \displaystyle\int_{-\alpha-\beta}^{-\beta} \dfrac{(-\beta - t)^{\frac{-1}{2}} dt}{z + t}, & \text{if } 0 < \alpha \le |\beta|, \beta < 0. \end{cases}$$

$F(2, w)$ has logarithmic branch points at $w = \pm i$. Moreover, $F(2, w)$ can be represented by a positive T-fraction

$$f(\alpha, \beta, 2, w) := \begin{cases} \dfrac{w\alpha^{\frac{1}{2}}}{2(\alpha - \beta w^2)} \left[\dfrac{F_1 \cdot (\frac{\alpha}{w^2} - \beta)}{1 + G_1 \cdot (\frac{\alpha}{w^2} - \beta)} + \dfrac{F_2 \cdot (\frac{\alpha}{w^2} - \beta)}{1 + G_2 \cdot (\frac{\alpha}{w^2} - \beta)} + \cdots \right], & \text{if } \alpha, \beta > 0, \\[4mm] \dfrac{w\alpha^{\frac{1}{2}}}{2(\alpha - \beta w^2)} \left[\dfrac{F_1 \cdot (\beta - \frac{\alpha}{w^2})}{1 + G_1 \cdot (\beta - \frac{\alpha}{w^2})} + \dfrac{F_2 \cdot (\beta - \frac{\alpha}{w^2})}{1 + G_2 \cdot (\beta - \frac{\alpha}{w^2})} + \cdots \right], & \text{if } 0 < \alpha \le |\beta|, \\ & \beta < 0, \end{cases}$$

$$(3.3)$$

and the positive coefficients F_l and G_l are computed by the FG-relations (1.8).

To illustrate the behavior of (3.3), we again consider contour maps of the number of significant digits in the approximations of $Arctan(w)$.

We note that for computation of $Arctan(w)$ it suffices to consider $w \in \mathbb{C}$ such that $|w| \le 1$ since $Arctan(w) + Arctan(\frac{1}{w}) = \frac{\pi}{2}$. Furthermore, it is only necessary to consider $|w| \le 1$ and $0 \le arg(w) \le \frac{\pi}{2}$ since $Arctan(w) = -Arctan(-w)$ and $Arctan(\overline{w}) = \overline{Arctan(w)}$.

To calculate $W = Arctan(w)$, we recall the continued fraction representation (1.15) and use Theorem 1.1 to determine which approximant of (1.15) will be used. As in Example 1. for $W = Log(1 + w)$, we use a representation for $W = Arctan(w)$ independent of the positive T-fraction. Then, using the truncation error theorem (Theorem 1.1) and the approximation formula for the number of significant digits (3.2), we have the inequality

$$\widetilde{SD}(f_m(w)) \ge -log_{10}\left|\frac{f_m(w) - f_{m-1}(w)}{f_m(w)}\right| - .301. \tag{3.4}$$

Finally, by considering the sets of points

$$S_1 := \left\{ w \in \mathbb{C} \mid w = \frac{j}{49} + i\frac{k}{49}, \quad j, k = 1, 2, 3, \cdots, 49 \right\},$$

$$S_2 := \left\{ w \in \mathbb{C} \mid w = \frac{j}{1000} + i\frac{k}{1000}, \quad j, k = 1, 2, 3, \cdots, 1000 \right\},$$

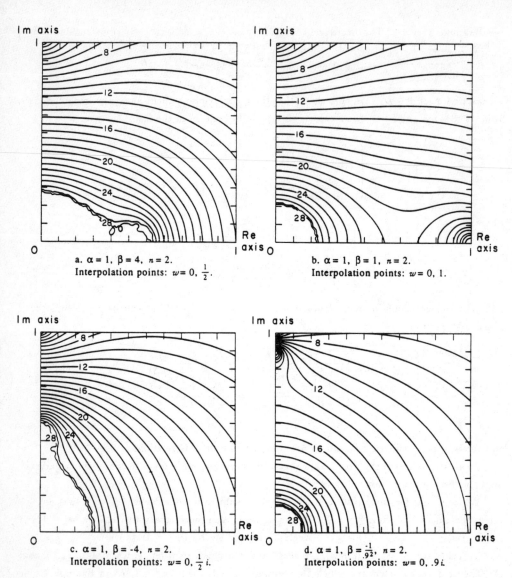

Figure 2. Level curves of the number $\widetilde{SD}(f_{10}(\alpha,\beta,2,w))$ of significant digits in the approximation of $F(2,w) = Arctan(w)$ by the positive T-fraction approximant $f_{10}(\alpha,\beta,2,w)$.

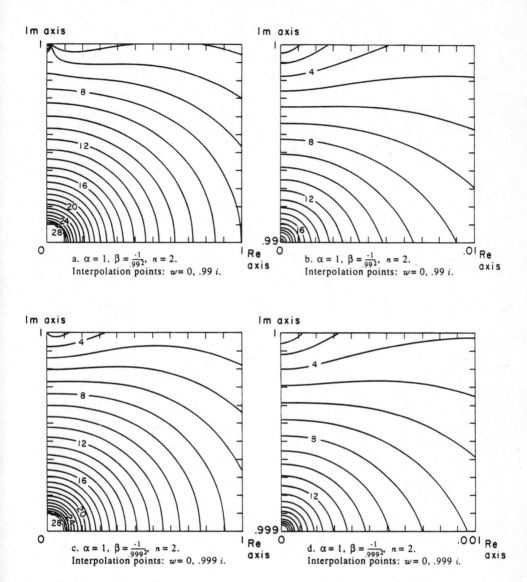

Figure 3. Level curves of the number $\widetilde{SD}(f_{10}(\alpha,\beta,2,w))$ of significant digits in the approximation of $F(2,w) = Arctan(w)$ by the positive T-fraction approximant $f_{10}(\alpha,\beta,2,w)$.

and

$$S_3 := \left\{ w \in \mathcal{C} \mid w = \frac{j}{10000} + i\frac{k}{10000}, \quad j, k = 1, 2, 3, \cdots, 10000 \right\},$$

we find that for each $w \in S_1$, S_2, and S_3, the values $m = 100$, $m = 300$, and $m = 900$, respectively, make the inequality of (3.4) greater than or equal to 28. Thus, for the following contour maps, the 100-th approximant of (1.15) is used for Figure 2, the 300-th approximant for Figures 3a and 3b, and the 900-th approximant for Figures 3c and 3d. Although the 100-th (300-th, 900-th) approximant of (1.15) is sufficient to yield 28 significant digits for all grid points used to plot the contour maps, such approximants were needed only for the grid points close to the singularity $w = i$.

Figure 2 gives level curves of $\widetilde{SD}(f_{10}(\alpha, \beta, 2, w))$ for various values of α and β in the region $\{w : 0 \le Re(w) \le 1, 0 \le Im(w) \le 1\}$. Each figure shows that approximations are best near the interpolation points $w = 0$ and $w = \sqrt{\frac{\alpha}{\beta}}$. The approximations are worst near the branch point $w = i$. In Figure 2a the interpolation points are at $w = 0$ and $w = \frac{1}{2}$. Keeping the interpolation point at $w = 0$ fixed, Figures 2b, 2c, and 2d have the other interpolation point at $w = 1, \frac{1}{2}i$, and $.9i$, respectively.

Table 2

a. $\alpha = 1$, $\beta = 4$.

b. $\alpha = 1$, $\beta = 1$.

K	MOMENTS	K	MOMENTS
-10	.10645875384043664667626107D-05	-10	.58247614038572620443330142D+00
-9	-.44486060450067285651122334D-05	-9	-.61650966334959245175290739D+00
-8	.18639385792028708544478863D-04	-8	.65708947423956528186976788D+00
-7	-.78323508027200590653139716D-04	-7	-.70643289533491645739821157D+00
-6	.33013894411869348648642785D-03	-6	.76781315854718159889895807D+00
-5	-.13961730849719710510507904D-02	-5	-.84618128727464621098773119D+00
-4	.59253626741575819476607563D-02	-4	.94920718545673852684312136D+00
-3	-.25241740835956393348771630D-01	-3	-.10890486225480862322117456D+01
-2	.10795595112510076452678203D+00	-2	.12853981633974483096156608D+01
-1	-.46364760900080611621425623D+00	-1	-.15707963267948966192313217D+01
0	.20000000000000000000000000D+01	0	.20000000000000000000000000D+01
1	-.86666666666666666666666667D+01	1	-.26666666666666666666666667D+01
2	.37733333333333333333333333D+02	2	.37333333333333333333333333D+01
3	-.16508571428571428571428571D+03	3	-.54857142857142857142857143D+01
4	.72586031746031746031746032D+03	4	.84317460317460317460317460D+01
5	-.32076738816738816738816739D+04	5	-.13483405483405483405483405D+02
6	.14247565101565101565101565D+05	6	.22292374292374292374292374D+02
7	-.63607576379176379176379176D+05	7	-.37872882672882672882672883D+02
8	.28541969930984048631107455D+06	8	.65762713103889574477809772D+02
9	-.12871825447530797375998614D+07	9	-.11619625451947433371581978D+03
10	.58336120752498275718089960D+07	10	.20818690906616603211030456D+03

L	F_l	L	F_l
1	.46364760900080611621425623D+00	1	.15707963267948966192313217D+01
2	.10167171141774411249355119D-02	2	.32911722786342361922106681D-01
3	.72485092203278575097240305D-03	3	.22146065141273590327084246D-01
4	.70768080229278667550837809D-03	4	.21734063378117205861897304D-01
5	.70249791541047034006300787D-03	5	.21601934268058519817404980D-01
6	.70026048862933236066881827D-03	6	.21543686417962275606844469D-01
7	.69909304859649393987462612D-03	7	.21512984467055920471931671D-01
8	.69840735071162902975890804D-03	8	.21494844334695838354377157D-01
9	.69792599089111317730593116D-03	9	.21483241163508314263974130D-01
10	.75346817897045723701546809D-03	10	.21475372872153774918860558D-01

L	G_l	L	G_l
1	.23182380450040305810712812D+00	1	.78539816339744830961566085D+00
2	.22248517994832220786722971D+00	2	.69731844029897070910483760D+00
3	.22328838647136210610027270D+00	3	.70412154697015079908844956D+00
4	.22345724557052491807509488D+00	4	.70568288750164597025938883D+00
5	.22351997987526389951291990D+00	5	.70627490079034294545851726D+00
6	.22355009103478688112272427D+00	6	.70656151591510765265024483D+00
7	.22356685370958785606536506D+00	7	.70672184169906569786709537D+00
8	.22357716226909502367058439D+00	8	.70682054175047011034463285D+00
9	.22355747185645758296520587D+00	9	.70688560364869310879194025D+00
10	.25947916101608514512300550D+00	10	.70693075053738571003845041D+00

Figure 3 gives the level curves of the number $\widetilde{SD}(f_{10}(\alpha,\beta,2,w))$ of significant digits in the approximation of $F(2,w) = Arctan(w)$ by the positive T-fraction approximant $f_{10}(\alpha,\beta,2,w)$ for other values of α and β. Figures 3a and 3b show the contour curves for $\alpha = 1$ and $\beta = \frac{-1}{.99^2}$ and have the interpolation points $w = 0$ and $w = .99i$. To see that the values of $\widetilde{SD}(f_{10}(1,\frac{-1}{.99^2},2,w))$ do increase as w approaches the interpolation point $w = .99i$, the region near $w = .99i$ is magnified in Figure 3b to illustrate the behavior. Figures 3c and 3d have the same relationship as above, except the interpolation point $w = .999i$ is now closer to the branch point $w = i$.

Table 2 contains the numerical values of the moments c_k, $k = -10, -9, \cdots, 9, 10$, and the positive continued fraction coefficient F_l and G_l, $l = 1, 2, \cdots, 10$ defined by (1.8) and (1.9) for values of α and β corresponding to those used in Figure 2. The value of $\alpha = 1$ is held fixed for each figure in Figure 2, and is therefore the same for Table 2. The β value for Figures 2a, 2b, 2c and 2d are 4, 1, -4 and $\frac{-1}{.9^2}$, respectively.

Table 2

c. $\alpha = 1$, $\beta = -4$.

d. $\alpha = 1$, $\beta = \frac{-1}{.9^2}$.

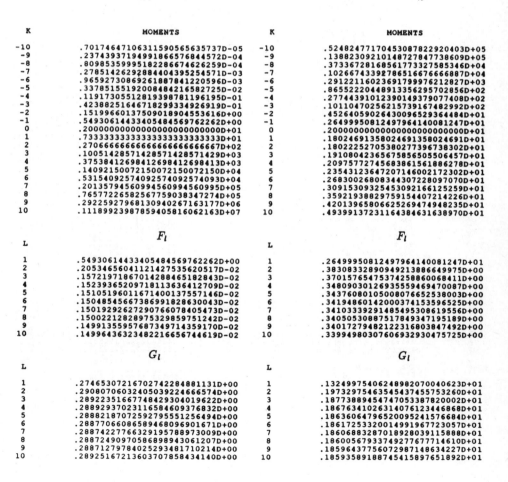

K	MOMENTS	K	MOMENTS
-10	.7017464710631159056563573 7D-05	-10	.5248247717045308782292040 3D+05
-9	.2374393719499186657684457 2D-04	-9	.1388230921014872784773860 9D+05
-8	.8098535999518228667462625 9D-04	-8	.3733672816856177332758534 6D+04
-7	.2785142629288440439525457 1D-03	-7	.1026674339278651667666688 7D+04
-6	.9659273086926188784122059 6D-03	-6	.2912211602369179997621282 7D+03
-5	.3378515519200848421658272 5D-02	-5	.8655222044891335629570285 6D+02
-4	.1191730551281939878119619 5D-01	-4	.2774439101239014937907740 8D+02
-3	.4238825164671829933492691 9D-01	-3	.1011047025621573916748299 2D+02
-2	.1519966013750901890455361 6D+00	-2	.4526405902643009652936448 4D+01
-1	.5493061443340548456976226 2D+00	-1	.2649995081249796414008124 7D+01
0	.2000000000000000000000000 0D+01	0	.2000000000000000000000000 0D+01
1	.7333333333333333333333333 3D+01	1	.1802469135802469135802469 1D+01
2	.2706666666666666666666666 7D+02	2	.1802225270538027739673830 2D+01
3	.1005142857142857142857142 9D+03	3	.1910804236567585650550645 7D+01
4	.3753841269841269841269841 3D+03	4	.2097577274568386156188627 8D+01
5	.1409215007215007215007215 0D+04	5	.2354312364720714600217230 2D+01
6	.5315409257409257409257409 3D+04	6	.2683002680834430722809707 0D+01
7	.2013579456099456099456099 5D+05	7	.3091530932545309216612525 9D+01
8	.7657722658256775903834727 4D+05	8	.3592193882975915440721422 6D+01
9	.2922592796813094026716317 7D+06	9	.4201396580662526947494823 5D+01
10	.1118992398785940581606216 3D+07	10	.4939913723116438463163897 0D+01

F_l

F_l

L		L	
1	.5493061443340548456976226 2D+00	1	.2649995081249796414008124 7D+01
2	.2053465604112142753562051 7D-02	2	.3830833289094921388664997 5D+00
3	.1572197186701428846518284 3D-02	3	.3701576547537425886006841 1D+00
4	.1523936520971811336412709 0D-02	4	.3480903012693555946947008 7D+00
5	.1510519601167140013755714 6D-02	5	.3437608010500807665253800 3D+00
6	.1504854566738699182863004 3D-02	6	.3419486014200037415359652 5D+00
7	.1501929262729076607840547 3D-02	7	.3410333929148549530861955 6D+00
8	.1500221282897532985975124 2D-02	8	.3405053088751784934719518 9D+00
9	.1499135595768734971435917 0D-02	9	.3401727948212231680384749 2D+00
10	.1499643632348221665674461 9D-02	10	.3399498030760693293047572 5D+00

G_l

G_l

L		L	
1	.2746530721670274228488113 1D+00	1	.1324997540624898207004062 3D+01
2	.2908070603240503922466657 4D+00	2	.1973297546354543745575326 0D+01
3	.2892235166774842930401962 2D+00	3	.1877388945474705338782000 2D+01
4	.2889293702311658460937683 2D+00	4	.1867634102631407612344686 8D+00
5	.2888218707259279555125649 4D+00	5	.1863606479652009524157668 4D+00
6	.2887706608658946809690167 1D+00	6	.1861725332001491967723057 0D+00
7	.2887422776632919578897300 9D+00	7	.1860688328701892803911588 8D+00
8	.2887249097058689894306120 7D+00	8	.1860056793374927767771161 0D+00
9	.2887127978402529348171021 4D+00	9	.1859643775607298714863422 7D+00
10	.2892516721360370785843414 0D+00	10	.1859358918874541589765189 2D+00

In Table 3. $\alpha = 1$ is again held fixed. Since Figures 3a and 3b have the same values for α and β. Table 3a,b shows the numerical values of the moments and positive continued fraction coefficients for the same values of α and β as in Figures 3a and 3b, that is, $\alpha = 1$ and $\beta = \frac{-1}{.99^2}$. Table 3c,d has the same relationship to Figures 3c and 3d, with the value of β moved to $\frac{-1}{.999^2}$.

Table 3

a,b. $\alpha = 1, \beta = \frac{-1}{.99^2}$.

c,d. $\alpha = 1, \beta = \frac{-1}{.999^2}$.

K	MOMENTS	K	MOMENTS
-10	.1896996338674367401698 8332D+15	-10	.21412901530976158835855826D+24
-9	.43339224624217842951203239D+13	-9	.48252257149566802194270066D+21
-8	.10059163870224416450003725D+12	-8	.11045916731422583946532989D+19
-7	.23836382407770188294153466D+10	-7	.25813380890601591241897225D+16
-6	.58107056068429187003377142D+08	-6	.62048278908719240452092578D+13
-5	.14760463922221313093311935D+07	-5	.15536669053327025919439813D+11
-4	.40031625957306547340468863D+05	-4	.41500313649874860472618268D+08
-3	.12260781697935449739534945D+04	-3	.12475199564258926870725514D+06
-2	.50839200470469710299266420D+02	-2	.50204093602308301564784693D+03
-1	.52403717764772474714560201D+01	-1	.75928019321658996634883826D+01
0	.20000000000000000000000000D+01	0	.20000000000000000000000000D+01
1	.13739414345474951535557596D+01	1	.13373393413433453473493514D+01
2	.11216353105609922323121038D+01	2	.10720160347301041588963184D+01
3	.98092443498387939700016986D+00	3	.92071139127411425147014410D+00
4	.88963663719284582968939462D+00	4	.82004940433983559229149260D+00
5	.82518169562406595832552320D+00	5	.74699269689185229237810502D+00
6	.77717190142498264154737476D+00	6	.69091285500501869679311650D+00
7	.74008819643931201146406939D+00	7	.64614363913923624950390720D+00
8	.71069645612566837731950692D+00	8	.60935328699889773770749313D+00
9	.68696192172990038295476508D+00	9	.57843835964831378339695142D+00
10	.66753336331941350825443 8181D+00	10	.55199711809460805807662032D+00

F_l

F_l

L		L	
1	.52403717764772474714560201D+01	1	.75928019321658996634883826D+01
2	.70812632911258058098439232D+01	2	.62324240729606896781904583D+02
3	.10897575079080421415433187D+02	3	.15296908403860565635563150D+03
4	.92956908352439223563674301D+01	4	.11161257792767532803111683D+03
5	.92646195047397965460792516D+01	5	.11843828434477723216029244D+03
6	.91783958161489745572791568D+01	6	.11443772892341117668337165D+03
7	.91497956128025328185034853D+01	7	.11511098738509700842659380D+03
8	.91305443387516047090212534D+01	8	.11442981465395956987779271D+03
9	.91190558525157942220223243D+01	9	.11443150579322075125198141D+03
10	.91112722095063802487129310D+01	10	.11426029060862595764722295D+03

G_l

G_l

L		L	
1	.26201858882386237357280100D+01	1	.37964009660829498317441913D+01
2	.88516789272773487586444623D+01	2	.40508738352257155395705829D+02
3	.69101714198079473173905942D+01	3	.19399881860671361026708305D+02
4	.70480449178979796480827974D+01	4	.23958054958257594695829360D+02
5	.69812213046154917638631012D+01	5	.21988447045502332377175302D+02
6	.69741322796822289406541944D+01	6	.22609852850829438513478211D+02
7	.69652456517717494592091031D+01	7	.22318212069387582376932555D+02
8	.69608423595836806047944669D+01	8	.22401671748475289233990357D+02
9	.69577666100679213130117812D+01	9	.22349527651693433938750469D+02
10	.69556887631710225433276908D+01	10	.22357332003887433491785350D+02

4. Relationship between the moments of $G(n, z)$ and incomplete beta function

To describe a relationship between the moments of $G(n,z)$ and a special function, the beta function, we start with a few definitions. The *beta function* is defined as

$$B(a, b) := \int_0^1 t^{a-1}(1-t)^{b-1}dt, Re(a) > 0, Re(b) > 0.$$

Note: $B(a, b) = \frac{\Gamma(a)\Gamma(b)}{\Gamma(a+b)} = B(b, a)$. The *complementary incomplete beta function* is defined as

$$C_z(a, b) := \int_z^1 t^{a-1}(1-t)^{b-1}dt, \quad Re(a) > 0, \quad Re(b) > 0, \quad 0 \le z \le 1.$$

Note: $C_0(a, b) = B(a, b)$.

Theorem 4.1 *If $0 \le \alpha \le 1, \beta = -1$, the moments*

$$c_k := (-1)^k \int_\beta^{\alpha+\beta} t^k(t-\beta)^{\frac{1}{n}-1}dt$$

of $G(n,z)$ in Theorem 2.1 can be expressed in terms of an incomplete beta function by

$$c_k = C_{1-\alpha}\left(k+1, \frac{1}{n}\right), \quad k = 1, 2, 3, \cdots.$$

Proof: Let $\tau = -t, d\tau = -dt$, and $\beta = -1$. Then

$$c_k = (-1)^k(-1)\int_1^{1-\alpha}(-1)^k\tau^k(-\tau-\beta)^{\frac{1}{n}-1}d\tau,$$

$$= -\int_1^{1-\alpha}\tau^k(1-\tau)^{\frac{1}{n}-1}d\tau,$$

$$= C_{1-\alpha}\left(k+1, \frac{1}{n}\right), \quad k = 1, 2, 3, \cdots.$$

Note: if $\alpha = 1$, then

$$c_k = C_0\left(k+1, \frac{1}{n}\right) = B\left(k+1, \frac{1}{n}\right) = \frac{\Gamma(k+1)\Gamma(\frac{1}{n})}{\Gamma(k+1+\frac{1}{n})}.$$

Recall that $\Gamma(\frac{1}{2}) = \sqrt{\pi}$. ∎

References

[1] M. Abramowitz and I.E. Stegun, eds., *Handbook of Mathematical Functions With Formulas, Graphs, and Mathematical Tables* (National Bureau of Standards, Washington, DC, 1964).

[2] E.T. Copson, *An Introduction to the Theory of Functions of a Complex Variable* (Oxford University Press, London, 1935).

[3] I.S. Gradshteyn and I.M Ryzhik, *Table of Integrals, Series, and Products* (Academic Press, New York, 1965).

[4] W.B. Gragg, Truncation error bounds for T-fractions, in: W. Cheney, ed., *Approximation Theory III* (Academic Press, New York, 1980) 455-460.

[5] P. Henrici, *Applied and Computational Complex Analysis, Volume 2: Special Functions, Integral Transforms, Asymptotics and Continued Fractions* (Wiley, New York, 1977).

[6] P. Henrici and P. Pluger, Truncation error estimates for Stieltjes fractions, *Numer. Math. 9* (1966) 120-138.

[7] T.H. Jefferson, Truncation error estimates for T-fractions, *SIAM J. Numer. Anal. 6* (1969) 359-364.

[8] W.B. Jones, Multiple point Padé tables, in: E.B. Saff and R.S. Varga, eds., *Padé and Rational Approximation* (Academic Press, New York, 1977) 163-171.

[9] W.B. Jones, O. Njåstad and W.J. Thron, Two-point Padé expansions for a family of analytic functions, *J Comp and Appl Math* 9 (1983) 105-123.

[10] W.B. Jones and W.J. Thron, *Continued Fractions: Analytic Theory and Applications*, Encyclopedia of Mathematics and its Applications 11 (Addison-Wesley, Reading, MA, 1980); (distributed by: Cambridge University Press, New York).

[11] W.B. Jones and W.J. Thron, Continued Fractions in Numerical Analysis, *Applied Numerical Mathematics* 4 (1988) 143-230.

[12] W.J. Thron, Some properties of Continued Fractions $1 + d_0 z + K(\frac{z}{1+d_n z})$, *Bull. Amer. Math. Soc.* 54 (1948) 206-218.

[13] N.J. Wyshinski, *Approximations for a Family of Stieltjes Transforms Associated with the Two-Point Padé Table*, M.S. Thesis, Mathematics Department, University of Colorado, Boulder, CO, 1988.

ON CONTINUED FRACTIONS ASSOCIATED WITH POLYNOMIAL TYPE PADE APPROXIMANTS, WITH AN APPLICATION

John H McCabe
Department of Mathematical Sciences
University of St Andrews
St Andrews, Fife KY16 9SS, U.K.

DEDICATED TO WOLF THRON ON THE OCCASION OF HIS 70TH BIRTHDAY

Abstract   The diagonal Padé approximants of power series that satisfy a certain reciprocal property involve, essentially, one polynomial. The continued fractions, whose convergents are the sequence of diagonal approximants, are consequentially of a simplified form.   An interesting example is a continued fraction given by Stieltjes, and this is seen to have an application in current approximation theory.

1.   Introduction

In a recent article [7] Nemeth and Zimani gave several examples of functions that admit power series expansions, possibly asymptotic, that have a particular reciprocal property.   Specifically, suppose that $f(x)$ is represented by the at least formal power series

$$f(x) = \sum_{k=0}^{\infty} c_k x^k \qquad (1)$$

and that

$$f(x).f(-x) = 1. \qquad (2)$$

Then the following results are simple consequences of (2).

(i)   The formal expansion of $\log_e\{f(x)\}$ contains only odd powers.

(ii)   The main diagonal Padé approximants $P_{n,n}(x)$ are of the form

$$P_{n,n}(x) = \frac{Q_n(x)}{Q_n(-x)}, \qquad n = 0,1,2,3,\ldots \qquad (3)$$

where $Q_n(x)$ is a polynomial of degree n.

(iii)   The regular C fraction corresponding to the power series, if it exists, has the form

$$f(x) = \frac{c_0}{1} + \frac{a_2 x}{1} - \frac{a_2 x/2}{1} + \frac{a_4 x}{1} - \frac{a_4 x}{1} + \frac{a_6 x}{1} - \frac{a_6 x}{1} + \ldots \, .$$

The first of these follows directly from (2). The second is easily established from the well known relationships between Padé approximants $P_{r,s}(x)$ for a normal power series and those for its reciprocal series. If the latter are denoted by $\tilde{P}_{r,s}(x)$ then

$$\tilde{P}_{r,s}(x) = \{P_{s,r}(x)\}^{-1}.$$

The third property is verified by taking the odd contraction of the regular C fraction for the power series, that is the associated continued fraction whose convergents are the main diagonal Padé approximants, and using the previous result. Details are given by McCabe [6] but the result appears originally in Perron [8], see §62, Satz 15. The converse of (iii), that is that the corresponding series of a continued fraction of the form (3) satisfies (2), was proved by Arne Magnus [5].

Nemeth and Zimani used the term "polynomial type Padé approximants" for the rational functions (3), since they are essentially one polynomial, and the resulting simplification in their evaluation is clear. The most well known series that satisfies (2) is of course the exponential series, for which the corresponding fraction is

$$e^x = \frac{1}{1} - \frac{x}{1} + \frac{x/2}{1} - \frac{x/6}{1} + \frac{x/6}{1} - \frac{x/10}{1} + \frac{x/10}{1} - \frac{x/14}{1} + \frac{x/14}{1} - \dots .$$

A second example is

$$f(x) = 1 + \frac{x}{12} + \frac{x^2}{288} - \frac{139x^3}{51840} - \dots .$$

The series $f(1/z)$ is Stirlings series in the asymptotic expansion for $\Gamma(z)$,

$$\Gamma(z) \sim (2\pi/z)^{1/2} z^z e^{-z} f(1/z).$$

The first twenty one coefficients of this series are given, as rational numbers, by Wrench [14], and the coefficients of the first ten polynomial Padé approximants are given by Nemeth and Zimani [7]. Further examples looked at in this latter paper are

$$\frac{\Gamma(x+\tfrac{1}{2})}{\Gamma(x+1)} \sim x^{-1/2}\left\{1 - \frac{1}{8x} + \frac{1}{128x^2} + \frac{5}{1024x^4} - \dots\right\}$$

and the series expansion of the binomial function $\{(1-x)/(1+x)\}^{\alpha}$. In this short article we look at two related examples of series that satisfies (2), and the continued fractions that correspond to them, and we will see that they can be used to provide error bounds for least squares approximation. One of these series has a particularly nice continued fraction expansion, and we begin with this expansion.

2. An expansion given by Stieltjes

In a letter to Hermite in 1891 Stieltjes, (see [11]), gave the following continued fraction for a ratio of gamma functions,

$$G(x) = x \left\{ \frac{\Gamma(x)}{\Gamma(x+\frac{1}{2})} \right\}^2 = 1 + \frac{2}{8x-1} + \frac{1.3}{8x} + \frac{3.5}{8x} + \frac{5.7}{8x} + \frac{7.9}{8x} + \cdots . \qquad (4)$$

This expansion is the odd part of the continued fraction

$$\frac{1}{1} - \frac{1/4x}{1} + \frac{1/8x}{1} - \frac{3/8x}{1} + \frac{3/8x}{1} - \frac{5/8x}{1} + \frac{5/8x}{1} - \frac{7/8x}{1} + \cdots \qquad (5)$$

which itself corresponds to the series

$$1 + \frac{1}{4x} + \frac{1}{32x^2} \quad \frac{1}{128x^3} \quad \frac{1}{2048x^4} + \cdots . \qquad (6)$$

This series is a particular case of a general asymptotic expansion for the ratio of two gamma functions given by Tricomi and Erdélyi [13]. Specifically,

$$\frac{\Gamma(z+a)}{\Gamma(z+b)} = \sum_{j=0}^{m-1} (-)^j \frac{\Gamma(b-a+j)}{\Gamma(b-a)j!} B_j^{(a-b+1)}(a) z^{a-b-j} + O(z^{a-b-m}),$$

valid for all $m \geq 1$ as $z \to \infty$ with $|\arg(z+a)| < \pi$, with $B_j^{(\alpha)}(a)$ being the generalised Bernoulli polynomials. Setting $a = 0$ and $b = 1/2$, and then squaring the result yields (6). For integer values of x the expansion (4) yields

$$\left(\frac{2.4.6 \ldots 2n}{1.3 \ldots (2n-1)} \right)^2 \frac{1}{n\pi} = 1 + \frac{2}{8n-1} + \frac{1.3}{8n} + \frac{3.5}{8n} + \frac{5.7}{8n} + \cdots .$$

Details of Stieltjes work can be found in Perron [8].

Setting $x = n/2$ in the left hand side of (4) and then (5) yields

$$(n/2) \left\{ \frac{\Gamma(n/2)}{\Gamma((n+1)/2)} \right\}^2 = \frac{1}{1} - \frac{1/2n}{1} + \frac{1/4n}{1} - \frac{3/4n}{1} + \frac{3/4n}{1} - \frac{5/4n}{1} + \frac{5/4n}{1} - \cdots$$

corresponding to the series

$$1 + \frac{1}{2n} + \frac{1}{8n^2} - \frac{1}{16n^3} - \frac{5}{128n^4} + \cdots .$$

The square root of this function

$$g(n) = \frac{(n/2)^{1/2}\Gamma(n/2)}{\Gamma((n+1)/2)}$$

has an asymptotic series which clearly satisfies $g(-n)g(n) = 1$. This particular gamma function ratio, together with others closely related to it, has been studied by many people from time to time, for instance Bowman and Shenton [1], who include some historic details. The series expansion for $g(n)$ begins

$$g(n) \sim 1 + \frac{1}{4n} + \frac{1}{32n^2} - \frac{5}{128n^3} - \frac{21}{2048n^4} + \dots \quad (7)$$

A recurrence relation for the coefficients of this series has been obtained by the present author [6]. The relation involves Bernoulli numbers and, as with apparently all recurrence formulae for these numbers, it involves a double summation. It is proved in [6] that each coefficient is an odd number divided by a power of two. Frenzen [3] has shown that the even order coefficients of the above series alternate in sign, and a similar method shows that the odd order coefficients do so as well. Frenzen [4] then used the asymptotic expansion to provide upper and lower bounds for the error of the best L_2 polynomial approximation of degree n for x^{n+1} on $[-1,1]$.
The odd contraction of the continued fraction

$$\frac{1}{1} + \frac{a_2 x}{1} + \frac{a_3 x}{1} + \frac{a_4 x}{1} + \frac{a_5 x}{1} + \dots$$

is

$$1 - \frac{a_2 x}{1 + (a_2 + a_3)x} - \frac{a_3 a_4 x^2}{1 + (a_4 + a_5)x} - \frac{a_5 a_6 x^2}{1 + (a_6 + a_7)x} - \dots \; .$$

If $a_3 = -a_2/2$ and

$$a_{2k+1} = -a_{2k}, \quad k = 2,3,4,\dots,$$

as is the case when the corresponding series satisfies (2), then this continued fraction becomes

$$1 - \frac{a_2 x}{1 + a_2 x/2} + \frac{a_2 a_4 x^2/2}{1} + \frac{a_4 a_6 x^2}{1} + \frac{a_6 a_8 x^2}{1} + \dots \; .$$

It follows that if $a_2, a_4, a_6, \dots$ are real and have the same sign and x is real then the convergents of this continued fraction form bounding sequences. This is certainly the case of the continued fraction corresponding to the exponential series, and for the continued fraction

(5) given by Stieltjes, the former being convergent, the latter asymptotic. The author has calculated the first sixty coefficients of the continued fraction that correspond to the series (7) and these are strictly alternating in sign, that is the even order ones are of the same sign. Clearly this alternation property is maintained if the exponential series is raised to any power α, where α is any real number. Numerical evidence suggests that the continued fractions corresponding to

$$\{G(x)\}^\alpha \sim \left\{1 + \frac{1}{4x} + \frac{1}{32x} - \frac{1}{128x^3} - \frac{5}{2084x^4} \cdots\right\}^\alpha$$

obtained from (4) and (6) and where α is any real number, also shares this property. Setting $\alpha = 1/2$, and $x = n/2$, yields (7). The author has not been able to obtain a condition (or conditions) in addition to $f(x)f(-x) = 1$ that imply

$$\text{sign } a_{2k} = \text{sign } a_2 \tag{8}$$

for all $k > 1$ as well as satisfying property (iii).

3. Least squares approximation

Phillips [9] has proved:

If $f \in C^{n+1}[-1,1]$ then there exists an $\xi \in (1,1)$ such that

$$\inf_{q \in P_n} \|f-q\|_2 = \delta_n |f^{(n+1)}(\xi)|/(n+1)!$$

where P_n denotes the set of all polynomials of degree n or less and where, following Timan [12], δ_n can be expressed as

$$\delta_n = \left(\frac{2}{2n+3}\right)^{1/2} 2^{n+1} \frac{\{(n+1)!\}^2}{(2n+2)!} .$$

Replacing the factorials by gamma functions and using the duplication formula yields

$$\delta_n = \frac{(2\pi)^{1/2}(2n+3)^{1/2}\Gamma(n+2)}{2^{n+2}\Gamma(n + 5/2)} .$$

Writing, as in the previous section,

$$g(n) = \frac{(n/2)^{1/2}\Gamma(n+2)}{\Gamma((n+1)/2)}$$

then gives, after some simplification,

$$\delta_n = \frac{\sqrt{\pi}}{2^{n+1}} \, g(-2n-3) \tag{9}$$

and hence

$$\delta_n^2 = \frac{\pi}{4^{n+1}} \, g^2(-2n-3). \tag{10}$$

Continued fraction expansions for δ_n and δ_n^2 can then be obtained from those of $g(n)$ and $g^2(n)$ by using the following result.

Let $F(n)$ denote the series expansion, possibly asymptotic or even merely formal,

$$F(n) = 1 + \frac{c_1}{n} + \frac{c_2}{n^2} + \frac{c_3}{n^3} + \dots + \frac{c_k}{n^k} + \dots$$

and let the corresponding (Stieltjes) fraction, assuming that it exists, be

$$F(n) = \frac{1}{1} + \frac{\alpha_2/n}{1} + \frac{\alpha_3/n}{1} + \frac{\alpha_4/n}{1} + \dots \; .$$

Suppose a and b are any real numbers, $a \neq 0$, and that

$$F(an+b) = 1 + \frac{d_1}{n} + \frac{d_2}{n^2} + \frac{d_3}{n^3} + \dots$$

and

$$F(an+b) = \frac{1}{1} + \frac{\beta_2/n}{1} + \frac{\beta_2/n}{1} + \frac{\beta_4/n}{1} + \dots \; .$$

Then $\beta_2 = \alpha_2/a$ and

$$\beta_{2s} + \beta_{2s+1} = \frac{b}{a} + \frac{\alpha_{2s} + \alpha_{2s+1}}{a}$$

and

$$\beta_{2s+1}\beta_{2s+2} = \frac{\alpha_{2s+1}\alpha_{2s+2}}{a^2}$$

for $s = 1,2,3,\dots$.

In the expansions for $g(n)$ and $g^2(n)$ the α's in each case satisfy $\alpha_{2s}+\alpha_{2s+1} = 0$ and so $\beta_{2s}+\beta_{2s+1} = b/a$ and $\beta_{2s+1}\beta_{2s+2} = -\alpha_{2s}\alpha_{2s+2}/a^2$.
Hence from the first of

$$g(n) = \frac{1}{1} - \frac{1/4n}{1} + \frac{1/8n}{1} - \frac{11/8n}{1} + \frac{11/8n}{1} - \frac{709/968n}{1} + \frac{709/968n}{1} - \ldots$$

and

$$g^2(n) = \frac{1}{1} - \frac{1/2n}{1} + \frac{1/4n}{1} - \frac{3/4n}{1} + \frac{3/4n}{1} - \frac{5/4n}{1} + \frac{5/4n}{1} - \ldots$$

we obtain

$$\delta_n = \frac{\sqrt{\pi}}{2^{n+1}} \left\{ \frac{1}{1} + \frac{1/8n}{1} + \frac{23/16n}{1} - \frac{11/368n}{1} + \ldots \right\}. \tag{11}$$

In the second we see that

$$\alpha_{2k+1} = -\alpha_{2k} = (2k-1)/4, \qquad k = 2,3,4,\ldots$$

and writing

$$\delta_n^2 = \frac{\pi}{4^{n+1}} \left\{ \frac{1}{1} + \frac{\beta_2/n}{1} + \frac{\beta_3/n}{1} + \frac{\beta_4/n}{1} + \ldots \right\}$$

yields

$$\beta_{2k} + \beta_{2k+1} = 3/2$$

and

$$\beta_{2k+1}\beta_{2k+2} = -\frac{(2k-1)(2k+1)}{64} = -\frac{(4k^2-1)}{64}$$

with $\beta_2 = 1/4$.
Hence

$$\delta_n^2 = \frac{\pi}{4^{n+1}} \left\{ \frac{1}{1} + \frac{1/4n}{1} + \frac{5/4n}{1} - \frac{3/16n}{1} + \frac{27/16n}{1} - \ldots \right\} \tag{12}$$

The odd contractions of (11) and (12) begin, respectively,

$$\delta_n = \frac{\sqrt{\pi}}{2^{n+1}} \left\{ 1 - \frac{1/8n}{1 + \frac{25}{16n}} + \frac{11/256n^2}{1 + \frac{3}{2n}} + \ldots \right\} \tag{13}$$

and

$$\delta_n^2 = \frac{\pi}{4^{n+1}} \left\{ 1 - \frac{1/4n}{1 + \frac{3}{2n}} + \frac{15/64n^2}{1 + \frac{3}{2n}} + \frac{35/64n^2}{1 + \frac{3}{2n}} + \frac{63/64n^2}{1 + \frac{3}{2n}} + \ldots \right\}. \tag{14}$$

The second and third convergents of (13) yield the bounds

$$\frac{1 + 47/16n + 563/256n^2}{1 + 49/16n + 611/256n^2} \cdot \frac{\sqrt{\pi}}{2^{n+1}} > \delta_n > \frac{1 + 23/16n}{1 + 25/16n} \cdot \frac{\sqrt{\pi}}{2^{n+1}}.$$

These bounds are sharper than

$$\frac{\sqrt{\pi}}{2^{n+1}} \left\{ 1 - \frac{1}{8n} + \frac{25}{128n^2} \right\} > \delta_n > \frac{\sqrt{\pi}}{2^{n+1}} \left\{ 1 - \frac{1}{8n} \right\}$$

established by Phillips and Sahney [10], which in turn are sharper than those given by the well known inequality

$$\frac{1}{2^{n+1/2}} < \delta_n < \frac{1}{2^{n-1/2}} .$$

Burgoyne [2] gives values of δ_n to five decimal places for n = 1 to n = 6.

While the partial denominators of (13) are all, after the first, 1+3/2n, the partial numerators do not follow a known pattern. The kth partial quotient of (14) however is, for $k \geq 2$, $\{(4k^2-1)64n^2\}/(1+3/2n)$ and so it is easy to obtain a sequence of bounds for δ_n^2.

References

1. Bowman, K O and Shenton, L R, 'Asymptotic series and Stieltjes continued fractions for a gamma function ratio', J. Comp. App. Maths., V.4, No.2, 1978, pp 105-111.

2. Burgoyne, F D, 'Practical L^p polynomial approximation', Maths. of Comp., V.21, 1967, pp 113-115.

3. Frenzen, C L, Private Communication.

4. Frenzen, C L, 'Error bounds for asymptotic expansions of the ratio of two gamma functions' Preprint of paper submitted to SIAM J. Math. Anal.

5. Magnus, A, Private Communication.

6. McCabe, J H, 'On an asymptotic series and corresponding continued fraction for a gamma function ratio', J. Comp. App. Maths., V.9, 1983, pp 125-130.

7. Németh, G and Zimani, M, 'Polynomial type Padé approximants', Maths. of Comp. V.38, No.158, 1982, pp 553-565.

8. Perron, O, 'Die Lehre von den Kettenbruchen', Chelsea, New York 1929.

9. Phillips, G M, 'Error estimates for best polynomial approximations', contributed to Approximation Theory' edited by A Talbot, Academic Press 1970.

10. Phillips, G M and Sahney, B N, 'An error estimate for least squares approximation', B.I.T., V.15, 1975, pp 426-430.

11. Stieltjes, T J, 'Correspondence d'Hermite et de Stieltjes', Tomes 1 and 2, Gauthier-Villars, 1905.

12.   Timan, A F, 'Theory of approximation of functions of a real var-
      iable', Pergamon Press, 1963 (Translated from Russian by J Berry).

13.   Tricomi, F C and Erdelyi, A, 'The asymptotic expansion of a ratio
      of gamma functions', Pacific J. Math., V.1, 1951, pp 133-142.

14.   Wrench, J W, 'Concerning two series for the gamma function',
      Maths. of Comp., V.22, 1968, pp 617-626.

MULTIPOINT PADE APPROXIMANTS

AND RELATED CONTINUED FRACTIONS

Olav Njåstad
Department of Mathematics
University of Trondheim – NTH
N–7034 Trondheim
NORWAY

1. Introduction.

Multipoint Padé fractions were introduced in [4]. They are continued fractions defined in the following way:

Let $\{a_1, a_2, \dots, a_p\}$ be given, fixed points in the complex plane. Every number $n \in N$ can be written in a unique way in the form $n = pq_n + r_n$, where $1 \le r_n \le p$. When there is no danger of confusion we write q for q_n, r for r_n. By a_{r-1} we mean a_p when $r = 1$, by a_{r+1} we mean a_1 when $r = p$, and so on. Statements relating to a_{r-1} when $r = 1$ and to a_{r+1} when $r = p$ etc., will be clear from the context.

Let A_n, B_n, C_n be constants, $B_n \neq 0$, $C_n \neq 0$ for all $n \in N$. We define

(1.1a) $\qquad a_1(z) = \dfrac{C_1}{z-a_1}, \ b_1(z) = \dfrac{A_1}{z-a_1} + B_1,$

(1.1b) $\qquad a_2(z) = \dfrac{C_2}{z-a_2}, \ b_2(z) = \dfrac{A_2(z-a_1)}{z-a_2} + \dfrac{B_2}{z-a_2},$

(1.1c) $\qquad a_n(z) = \dfrac{C_n(z-a_{r-2})}{z-a_r}, \ b_n(z) = \dfrac{A_n(z-a_{r-1}) + B_n(z-a_{r-2})}{z-a_r}$

$$\text{for } n = 3, 4, \dots \ .$$

The continued fraction $\overset{\infty}{\underset{n=1}{K}} \dfrac{a_n(z)}{b_n(z)}$ is then called a multipoint Padé continued fraction, or MP–fraction (belonging to the set $\{a_1, a_2, ..., a_p\}$).

In [4] it was shown that a linear functional ϕ on the space of R–functions belonging to $\{a_1, ..., a_p\}$ with associated regular orthogonal R–functions, gives rise to an MP–fraction. Furthermore, every MP–fraction whose denominators are of exact degree (and hence regular) originates in this way. (R–functions belonging to the points $\{a_1, ..., a_p\}$ are rational functions with no poles in the extended complex plane outside $\{a_1, ..., a_p\}$. For more information on the theory of functionals on the space of R–functions, see [8,9], where in particular the concept of orthogonal R–functions is defined. For the concepts of regular R–functions and R–functions of exact degree, see [4,8,9].) It follows from the results of [10,11] together with [4], that the nth approximant of an MP–fraction with denominators of exact degree is a multipoint Padé approximant of type $(n-1,n)$ for formal power series defined by the moments of the functional ϕ determined by the MP–fraction. For general discussions of multipoint Padé approximants, see e.g., [1,2,3,7,12,13].

In this paper we treat the relationship of MP–fractions to multipoint Padé approximants without reference to the theory of R–functions or functionals and their moments. We establish by direct arguments that every MP–fraction whose denominators are of exact degree (for definition, see Section 2) defines a system of formal power series

$$(1.2) \qquad L_0 = \frac{c_0}{z}, \; L_i = -\sum_{j=1}^{\infty} c_j^{(i)}(z-a_i)^{j-1}, \; i = 1,...,p,$$

such that the approximants of the MP–fraction are multipoint Padé approximants of type $(n-1,n)$ for these series. We also show by a direct argument that these multipoint Padé approximants are unique.

We wish to mention that in the case of <u>positive</u> MP–fractions, special results on the structure of the approximants can be proved. We refer to [5] for a discussion of these fractions.

For standard concepts and results on continued fractions we refer to [6].

2. MP–fractions and corresponding power series.

Let an MP–fraction $\overset{\infty}{\underset{n=1}{K}} \dfrac{a_n(z)}{b_n(z)}$ be given, with $a_n(z)$, $b_n(z)$ defined by (1.1), and let $P_n(z)/Q_n(z)$ be the nth approximant. Then the denominators $Q_n(z)$ satisfy the recurrence relations

$$(2.1a) \qquad Q_1 = \left[\frac{A_1}{z-a_1} + B_1 \right] Q_0 + \frac{C_1}{z-a_1} Q_{-1}$$

$$(2.1b) \qquad Q_2 = \frac{A_n(z-a_1) + B_2}{z-a_2} Q_1 + \frac{C_2}{z-a_2} Q_0$$

$$(2.1c) \qquad Q_n = \frac{A_n(z-a_{r-1}) + B_n(z-a_{r-2})}{z-a_r} Q_{n-1} + \frac{C_n(z-a_{r-2})}{z-a_r} Q_{n-2} \quad \text{for } n = 3,4,...,$$

with initial conditions $Q_{-1} = 0$, $Q_0 = 1$. The numerators $P_n(z)$ satisfy the same recurrence relation, with initial conditions $P_{-1} = 1$, $P_0 = 0$. It is easily verified by induction that we may write

$$(2.2) \qquad Q_n(z) = \beta_0^{(n)} + \frac{\beta_1^{(n)}}{(z-a_1)} + ... + \frac{\beta_p^{(n)}}{(z-a_p)} + .. + \frac{\beta_{n-1}^{(n)}}{(z-a_{r-1})^{q+1}} + \frac{\beta_n^{(n)}}{(z-a_r)^{q+1}},$$

$$(2.3) \qquad P_n(z) = \frac{\alpha_1^{(n)}}{(z-a_1)} + ... + \frac{\alpha_p^{(n)}}{(z-a_p)} + ... + \frac{\alpha_{n-1}^{(n)}}{(z-a_{r-1})^{q+1}} + \frac{\alpha_n^{(n)}}{(z-a_r)^{q+1}}.$$

Equivalently $Q_n(z)$ and $P_n(z)$ may be written in the following way:

$$(2.4) \qquad Q_n(z) = \frac{V_n(z)}{N_n(z)}, \quad P_n(z) = \frac{U_n(z)}{N_n(z)},$$

where

$$(2.5) \qquad N_n(z) = (z-a_1)^{q+1} \ ... \ (z-a_r)^{q+1}(z-a_{r+1})^q \ ... \ (z-a_p)^q,$$

and $U_n(z)$ and $V_n(z)$ are polynomials such that deg $V_n \leq n$, deg $U_n \leq n-1$. We shall say that $Q_n(z)$ is <u>of exact degree</u> if $\beta_n^{(n)} \neq 0$. This is equivalent to a_r not being a zero of $V_n(z)$. We say that $Q_n(z)$ is <u>regular</u> if $\beta_{n-1}^{(n)} \neq 0$. This is equivalent to a_{r-1} not being a zero of $V_n(z)$. It follows from the recurrence relation (2.1) that if $Q_n(z)$ is of exact degree, then $Q_n(z)$ is also regular, since $B_n \neq 0$. We say that $Q_n(z)$ is <u>degenerate</u> if deg $V_n < n$.

Lemma 2.1. Assume that all Q_n are of exact degree. Then the polynomials V_n and V_{n-1} do not both have a zero at a_i for any fixed $i = 1,...,p$. Similarly V_n and V_{n-1} are not both degenerate.

Proof: This lemma can be found in [5]. For the sake of completeness we give the proof.

The recurrence relations (2.1) can be rewritten in the following form:

$$(2.6a) \qquad V_1 = A_1 + B_1(z-a_1)$$

$$(2.6b) \qquad V_2 = \left[A_2(z-a_1) + B_2\right]V_1 + C_2(z-a_1)$$

$$(2.6c) \qquad V_n = \left[A_n(z-a_{r-1}) + B_n(z-a_{r-2})\right]V_{n-1} + C_n(z-a_{r-1})(z-a_{r-2})\, V_{n-2}$$
$$\text{for } n = 3,4,... \ .$$

Let $n \geq 2$, and assume that V_n and V_{n-1} have a common factor $(z-a_i)$. It follows from (2.6c) (with n replaced by $n+1$) that $(z-a_i)$ is also a factor of V_{n+1}. By repeating this argument at most $p-2$ times, we conclude that $(z-a_i)$ is a factor of V_{pq+i} for some q. This contradicts the assumption that all Q_n are of exact degree.

The proof of the second statement is similar: If $\deg V_n < n$, $\deg V_{n-1} < n-1$, then $\deg V_{n-2} < n-2$, and by repeating the argument we get $\deg V_1 < 1$, which is impossible since $B_1 \neq 0$.

∎

Theorem 2.2. Let an MP-fraction $\underset{n=1}{\overset{\infty}{K}} \dfrac{a_n(z)}{b_n(z)}$ be given, with approximants $\dfrac{P_n(z)}{Q_n(z)} = \dfrac{U_n(z)}{V_n(z)}$, where U_n, V_n, P_n, Q_n are as defined above. Assume that all Q_n are of exact degree. Then there exist a constant c_0 and sequences $\{c_j^{(i)}: j = 1,2,...\}$, $i = 1,...,p$, such that $\dfrac{U_n(z)}{V_n(z)}$ is a multipoint Padé approximant for the series $\dfrac{c_0}{z}, -\sum_{j=0}^{\infty} c_{j+1}^{(i)}(z-a_i)^j$, $i = 1,...,p$, in the following sense:

$$(2.7a) \qquad U_n(z) + V_n(z) \sum_{j=0}^{2q+1} c_{j+1}^{(i)}(z-a_i)^j = \sum_{j=2q+2}^{\infty} \gamma_j^{(i)}(z-a_i)^j \quad \text{when } i = 1,...,r-1,$$

$$(2.7b) \qquad U_n(z) + V_n(z) \sum_{j=0}^{2q} c_{j+1}^{(r)} (z-a_r)^j = \sum_{j=2q+1}^{\infty} \gamma_j^{(r)}(z-a_r)^j,$$

$$(2.7c) \qquad U_n(z) + V_n(z) \sum_{j=0}^{2q-1} c_{j+1}^{(i)}(z-a_i)^j = \sum_{j=2q}^{\infty} \gamma_j^{(r)}(z-a_r)^j \quad \text{when } i = r+1,...,p,$$

$$(2.7d) \qquad U_n(z) - V_n(z) \cdot \frac{c_0}{z} = \sum_{j=2}^{\infty} \gamma_{n-j}^{(0)} z^{n-j}.$$

Proof: From the product formula for continued fractions (see. e.g. [6]) we get

$$P_n(z)Q_{n-1}(z) - P_{n-1}(z)Q_n(z) = (-1)^{n-1} a_1(z)a_2(z)...a_n(z)$$

$$= (-1)^{n-1} \frac{C_1}{z-a_1} \cdot \frac{C_2}{z-a_2} \cdot \frac{C_3(z-a_1)}{z-a_3} \cdot \frac{C_n(z-a_{r-2})}{z-a_r}$$

and hence

$$(2.8) \qquad P_n(z)Q_{n-1}(z) - P_{n-1}(z)Q_n(z) = (-1)^{n-1} \frac{C_1 C_2...C_n}{(z-a_{r-1})(z-a_r)}.$$

Multiplication of the recurrence relations (2.1) by $P_{n-2}(z)$ and the same relation for $P_n(z)$ by $Q_{n-1}(z)$ and subtraction leads to

$$P_n(z)Q_{n-2}(z) - P_{n-1}(z)Q_n(z) = b_n(z)\left[P_{n-1}(z)Q_{n-2}(z) - P_{n-2}(z)Q_{n-1}(z)\right].$$

So again taking into account the product formula, we get

$$(2.9) \qquad P_n(z)Q_{n-2}(z) - P_{n-2}(z)Q_n(z) = (-1)^{n-2} \frac{C_1 C_2...C_n[A_n(z-a_{r-1})+B_n(z-a_{r-2})]}{(z-a_{r-2})(z-a_{r-1})(z-a_r)}.$$

We set $D_n(z) = \dfrac{P_n(z)}{Q_n(z)} - \dfrac{P_{n-1}(z)}{Q_{n-1}(z)}$, $E_n(z) = \dfrac{P_n(z)}{Q_n(z)} - \dfrac{P_{n-2}(z)}{Q_{n-2}(z)}$. From (2.8) and (2.9) we then obtain

$$(2.10) \qquad D_n(z) = (-1)^{n-1} \frac{C_1 C_2 \cdots C_n N_{n-1}(z)N_{n-2}(z)}{V_n(z)V_{n-1}(z)}$$

$$(2.11) \qquad E_n(z) = (-1)^{n-2} \frac{C_1 C_2...C_{n-1}[A_n(z-a_{r-1})+B_n(z-a_{r-2})]N_{n-2}(z)N_{n-3}(z)}{V_n(z)V_{n-2}(z)}.$$

Let i be fixed, and assume that $(z-a_i)$ is not a factor of $V_n(z)$. We define $c_1^{(i,n)},\ldots,c_{s_i}^{(i,n)}$ as the coefficients of $1, (z-a_i),\ldots,(z-a_i)^{s_i-1}$ in the power series expansion of $-\dfrac{P_n(z)}{Q_n(z)}$ about the point a_i. Here $s_i = 2q+2$ if $i < r$, $s_i = 2q+1$ if $i = r$, $s_i = 2q$ if $i > r$. For each n such that deg $V_n = n$ we define $c_0^{(n)}$ as the coefficient of z^{-1} in the power series expansion of $\dfrac{P_n(z)}{Q_n(z)}$ about ∞. (Recall that the constant term is zero, since deg $U_n < n-1$, deg $V_n = n$.

It follows immediately that series expansions of the following form are valid in the situations described:

$$(2.12a) \qquad U_n(z) + V_n(z) \sum_{j=0}^{2q+1} c_{j+1}^{(i,n)}(z-a_i)^j = \alpha_i(z-a_i)^{2q+2} + \ldots \text{ if } i < r,$$

$$(2.12b) \qquad U_n(z) + V_n(z) \sum_{j=0}^{2q} c_{j+1}^{(i,n)}(z-a_i)^j = \alpha_i(z-a_i)^{2q+1} + \ldots \text{ if } i = r,$$

$$(2.12c) \qquad U_n(z) + V_n(z) \sum_{j=0}^{2q-1} c_{j+1}^{(i,n)}(z-a_i)^j = \alpha_i(z-a_i)^{2q} + \ldots \text{ if } i > r,$$

$$(2.12d) \qquad U_n(z) - V_n(z) \frac{c_0^{(n)}}{z} = \alpha_0 z^{n-2} + \ldots \text{ (decreasing powers).}$$

As before, let i be fixed and $(z-a_i)$ not a factor of $V_n(z)$. Let m_i denote the multiplicity of $(z-a_i)$ as a factor in $V_{n-1}(z)$. Similarly set $m_0 = n-1-$ deg V_{n-1}, when deg $V_n = n$. From (2.9) it follows that power series expansions of the following form are valid:

$$(2.13a) \qquad D_n(z) = \beta_i(z-a_i)^{2q+2-m_i} + \ldots \text{ if } i < r-1,$$

$$(2.13b) \qquad D_n(z) = \beta_i(z-a_i)^{2q+1-m_i} + \ldots \text{ if } i = r-1,$$

$$(2.13c) \qquad D_n(z) = \beta_i(z-a_i)^{2q-m_i} + \ldots \text{ if } i > r-1,$$

$$(2.13d) \qquad D_n(z) = \beta_0 z^{m_0-2} + \ldots \text{ (decreasing powers).}$$

Assume that $i < r-1$. It follows from (2.13a) and the definition of $c_j^{(i,n)}$ that we may write

$$\frac{U_{n-1}(z)}{V_{n-1}(z)} + \sum_{j=0}^{2q+1} c_{j+1}^{(i,n)}(z-a_i)^j$$

$$= \left[\frac{U_{n-1}(z)}{V_{n-1}(z)} - \frac{U_n(z)}{V_n(z)}\right] + \left[\frac{U_n(z)}{V_n(z)} + \sum_{j=0}^{2q+1} c_{j+1}^{(i,n)}(z-a_i)^j\right]$$

$$= \delta_i(z-a_i)^{2q+2-m_i} + \dots .$$

It follows that

$$(2.14a) \qquad U_{n-1}(z) + V_{n-1}(z) \sum_{j=0}^{2q+1} c_{j+1}^{(i,n)}(z-a_i)^j = \varepsilon_i(z-a_i)^{2q+2} + \dots .$$

Next assume that $i = r-1$. It follows from (2.13b) and the definition of $c_j^{(r-1,n)}$ that we may write

$$\frac{U_{n-1}(z)}{V_{n-1}(z)} + \sum_{j=0}^{2q} c_j^{(r-1,n)}(z-a_{r-1})^j$$

$$= \left[\frac{U_{n-1}(z)}{V_{n-1}(z)} - \frac{U_n(z)}{V_n(z)}\right] + \left[\frac{U_n(z)}{V_n(z)} + \sum_{j=0}^{2q} c_{j+1}^{(r-1,n)}(z-a_{r-1})^j\right]$$

$$= \delta_{r-1}(z-a_{r-1})^{2q+1-m_{r-1}} + \dots .$$

As above we conclude that we may write

$$(2.14b) \qquad U_{n-1}(z) + V_{n-1}(z) \sum_{j=0}^{2q} c_{j+1}^{(r-1,n)}(z-a_{r-1})^j = \varepsilon_{r-1}(z-a_{r-1})^{2q+1} + \dots .$$

Next assume that $i > r-1$. It follows from (2.13c) and the definition of $c_j^{(i,n)}$ that we may write

$$\frac{U_{n-1}(z)}{V_{n-1}(z)} + \sum_{j=0}^{2q-1} c_{j+1}^{(i,n)}(z-a_i)^j$$

$$= \left[\frac{U_{n-1}(z)}{V_{n-1}(z)} - \frac{U_n(z)}{V_n(z)}\right] + \left[\frac{U_n(z)}{V_n(z)} + \sum_{j=0}^{2q-1} c_{j+1}^{(i,n)}(z-a_i)^j\right]$$

$$= \delta_i(z-a_i)^{2q-m_i} + \dots .$$

As above we conclude that we may write

(2.14c) $\quad U_{n-1}(z) + V_{n-1}(z) \sum_{j=0}^{2q-1} c_{j+1}^{(i,n)}(z-a_i)^j = \varepsilon_i(z-a_i)^{2q} + \dots .$

Similarly we conclude from (2.13d) and the definition of $c_0^{(n)}$ that we may write

$$\frac{U_{n-1}(z)}{V_{n-1}(z)} - \frac{c_0^{(n)}}{z} = \left[\frac{U_{n-1}(z)}{V_{n-1}(z)} - \frac{U_n(z)}{V_n(z)}\right] + \left[\frac{U_n(z)}{V_n(z)} - \frac{c_0^{(n)}}{z}\right]$$

$$= \beta_0 z^{m_0-2} + \dots \text{ (decreasing powers)}.$$

Taking into account that $m_0 = n - 1 - \deg V_{n-1}$, we conclude that we may write

(2.14d) $\quad U_{n-1}(z) - V_{n-1}(z)\dfrac{c_0^{(n)}}{z} = \varepsilon_0 z^{n-3} + \dots \text{ (decreasing powers)}.$

Now suppose that $(z-a_i)$ in addition to not being a factor of $V_n(z)$ also is not a factor of $V_{n-1}(z)$. It then follows from (2.13a) – (2.13c) that $c_j^{(i,n)} = c_j^{(i,n-1)}$ for $i = 1,\dots,$ σ_i, where $\sigma_i = 2q+2$ if $i > r-1$, $\sigma_i = 2q+1$ if $i = r-1$, $\sigma_i = 2q$ if $i > r-1$. Similarly suppose that in addition to $V_n(z)$ being of degree n also $V_{n-1}(z)$ is of degree n–1. It then follows from (2.13d) that $c_0^{(n)} = c_0^{(n-1)}$.

Next suppose that $(z-a_i)$ is a factor of $V_{n-1}(z)$. Then by Lemma 2.1, $(z-a_i)$ is not a factor of $V_{n-2}(z)$. From (2.11) it then follows that power series expansions of the following form are valid:

(2.15a) $\quad E_n(z) = \zeta_i(z-a_i)^{2q+2} + \dots \text{ if } i < r-2$

(2.15b) $E_n(z) = \zeta_i(z-a_i)^{2q+1} + \ldots$ if $i = r-2$

(2.15c) $E_n(z) = \zeta_i(z-a_i)^{2q} + \ldots$ if $i > r-2$.

Similarly assume that deg $V_n = n$ while deg $V_{n-1} < n-1$. Then by Lemma 2.1, deg $V_{n-2} = n-2$. From (2.11) it then follows that a power series expansion of the following form is valid:

(2.15d) $E_n(z) = \zeta_0 z^{-2} + \ldots$ (decreasing powers).

It follows from (2.15a)–(2.15c) that $c_j^{(i,n)} = c_j^{(i,n-2)}$ for $j = 1,\ldots, \tau_i$ if $(z-a_i)$ is a factor of $V_{n-1}(z)$, where $\tau_i = 2q+2$ if $i < r-2$, $\tau_i = 2q+1$ if $i = r-2$, $\tau_i = 2q$ if $i > r-2$. Similarly it follows from (2.15d) that $c_0^{(n)} = c_0^{(n-2)}$ if deg $V_{n-1} < n-1$.

Now let i and j be given, and let n be such that $(z-a_i)$ is not a factor of $V_n(z)$ and so large that $j \leq s_i$ (where s_i is defined as above when n is given). We define the constant $c_j^{(i)}$ by $c_j^{(i,n)}$. Similarly we define $c_0 = c_0^{(n)}$ for arbitrary n where deg $V_n = n$. It follows from the foregoing that $c_0, c_j^{(i)}$, $i = 1,\ldots,p, j = 1,2,\ldots$, are uniquely determinded in this way. Furthermore, Eqs. (2.12) and (2.14) imply that with these constants as series coefficients, Eqs. (2.7) are satisfied.

■

Remark. We note that if $(z-a_i)$ is not a factor of $V_n(z)$ and deg $V_n = n$, then $\dfrac{P_n(z)}{Q_n(z)}$ is a strong multipoint Padé approximant in the sense that Eq. (2.7) can be written

(2.16a) $\dfrac{U_n(z)}{V_n(z)} + \sum_{j=0}^{2q+1} c_{j+1}^{(i)}(z-a_i)^j = \sum_{j=2q+2}^{\infty} \kappa_j^{(i)}(z-a_i)^j$ when $i = 1,\ldots,r-1$,

(2.16b) $\dfrac{U_n(z)}{V_n(z)} + \sum_{j=0}^{2q} c_{j+1}^{(r)}(z-a_r)^j = \sum_{j=2q+1}^{\infty} \kappa_j^{(r)}(z-a_r)^j$,

(2.16c) $\dfrac{U_n(z)}{V_n(z)} + \sum_{j=0}^{2q-1} c_{j+1}^{(i)}(z-a_i)^j = \sum_{j=2q}^{\infty} \kappa_j^{(i)}(z-a_i)^j$ when $i = r+1,\ldots,p$,

(2.16d) $\dfrac{U_n(z)}{V_n(z)} - \dfrac{c_0}{z} = \sum_{j=2}^{\infty} \kappa_j^{(0)} z^{-j}$.

3. Uniqueness of approximants.

Theorem 3.1. There exists only one rational function $\dfrac{U_n(z)}{V_n(z)}$, where deg $U_n \leq n-1$, deg $V_n \leq n$, $V_n(z) \neq 0$, such that the Eqs. (2.7) are satisfied.

Proof: Let both $\dfrac{U_n(z)}{V_n(z)}$ and $\dfrac{u_n(z)}{v_n(z)}$ satisfy the conditions of the theorem. Then

$$(3.1) \qquad U_n(z) + V_n(z) \sum_{j=0}^{s_i-1} c_{j+1}^{(i)}(z-a_i)^j = \mu_i(z-a_i)^{s_i} + \ldots \; ,$$

where $s_i = 2q+2$ when $i < r, s_r = 2q+1$, $s_i = 2q$ when $i > r$.

Similarly

$$(3.2) \qquad u_n(z) + v_n(z) \sum_{j=0}^{s_i-1} c_{j+1}^{(i)}(z-a_i)^j = v_i(z-a_i)^{s_i} + \ldots \; ,$$

where s_i has the same meaning as above.

We also have

$$(3.3) \qquad U_n(z) - V_n(z) \frac{c_0}{z} = \mu_0 z^{n-2} + \ldots \text{ (decreasing powers)}.$$

and

$$(3.4) \qquad u_n(z) - v_n(z) \frac{c_0}{z} = v_0 z^{n-2} + \ldots \text{ (decreasing powers)}.$$

We define the expression $\pi(z)$ by

$$(3.5) \qquad \pi(z) = U_n(z)v_n(z) - u_n(z)V_n(z).$$

Then $\pi(z)$ is a polynomial of degree at most $2n-1$. Multiplication of (3.1) by $v_n(z)$ and (3.2) by $V_n(z)$ and subtraction gives for $i = 1,\ldots,p$:

$$(3.6) \qquad \pi(z) = \omega_i(z-a_i)^{s_i} + \ldots \text{ (increasing powers)}.$$

Similarly form (3.3) and (3.4) we get

$$(3.7) \qquad \pi(z) = \omega_0 z^{2n-2} + \dots \text{ (decreasing powers).}$$

Equation (3.7) shows that deg $\pi \leq 2n-2$. We conclude from Eq. (3.6) that $\pi(z)$ has a zero at a_i of multiplicity at least s_i. The sum of the multiplicities of these zeros is then
$$s_1 + \dots + s_{r-1} + s_r + s_{r+1} + \dots + s_p = (r-1)(2q+2) + (2q+1)+(p-r-1)(2q) = 2n-1.$$ It follows that $\pi(z) \equiv 0$, and so $\dfrac{U_n(z)}{V_n(z)} = \dfrac{u_n(z)}{v_n(z)}$.

∎

Remark. The argument does not show that $U_n(z) \equiv u_n(z)$, $V_n(z) \equiv v_n(z)$. It is possible that, e.g. $U_n(z)$ and $V_n(z)$ have a common factor.

REFERENCES

1. G.A. Baker, Jr. and P. Graves–Morris, Padé Approximants I, II, Encyclopedia of Mathematics and its Applications 13,14, Addison–Wesley Publ. Co. (1980).

2. C.K. Chui, "Recent results on Padé approximants and related problems", Approximation Theory II (Eds.: C.C. Lorentz, C.K. Chui and L.L. Schumaker), Academic Press (1976), 79–115.

3. M.A. Galucci and W.B. Jones, "Rational approximations corresponding to Newton series (Newton–Padé approximants)," J. Approximation Theory 17 (1976), 366–392.

4. E. Hendriksen and O. Njåstad, "A Favard type theorem for rational functions," J. Math. Anal. Appl., to appear.

5. E. Hendriksen and O. Njåstad, "Positive multipoint Padé continued fractions," Proc. Edinburgh Math. Soc. 32 (1989), 261–269.

6. W.B. Jones and W.J. Thron, Continued Fractions: Analytic Theory and Applications, Encyclopedia of Mathematics and its Applications 12, Addison–Wesley Publ. Co. (1980).

7. J. Karlsson, "Rational interpolation and best rational approximation," J. Math. Anal Appl. 52 (1976), 38–52.

8. O. Njåstad, "An extended Hamburger moment problem," Proc. Edinburgh Math. Soc. (Series II) 28 (1985), 167–183.

9. O. Njåstad, "Unique solvability of an extended Hamburger moment problem," J. Math. Anal. Appl. 24 (1987), 502–519.

10. O. Njåstad, "A multipoint Padé approximation problem," Analytic Theory of Continued Fractions II (Ed.: W.J. Thron), Springer Lecture Notes in Mathematics 1199 (1986), 263–268.

11. O. Njåstad, "Multipoint Padé approximation and orthogonal rational functions," Nonlinear Numerical Methods and Rational Approximation, (Ed.: A. Cuyt), Reidel Publishing Co. (1988), 259–270.

12. E.B. Saff, "An extension of Montessus de Ballore's theorem on the convergence of interpolation rational functions," J. Approximation Theory 6 (1972), 63–67.

13. H. Wallin, "Rational interpolation to meromorphic functions," Padé Approximation and its Applications (Eds.: M.G. de Bruin and H. van Rossum), Springer Lecture Notes in Mathematics 888 (1981), 371–382.

A survey of some results on separate

convergence of continued fractions.

Olav Njåstad

Department of Mathematics

University of Trondheim – NTH

N – 7034 Trondheim

Norway

0. Introduction.

Let $b_0 + \underset{n=1}{\overset{\infty}{K}} \dfrac{a_n(z)}{b_n(z)}$ be a continued fraction, where the elements in general are functions of a complex variable z. We denote the approximants by $f_n(z) = \dfrac{A_n(z)}{B_n(z)}$. The numerators $A_n(z)$ and denominators $B_n(z)$ satisfy the recurrence relations

$$\begin{bmatrix} A_n \\ B_n \end{bmatrix} = b_n \begin{bmatrix} A_{n-1} \\ B_{n-1} \end{bmatrix} + a_n \begin{bmatrix} A_{n-2} \\ B_{n-2} \end{bmatrix} , \quad n \geq 1,$$

$$A_{-1} = 1, \; A_0 = b_0, \; B_{-1} = 0, \; B_0 = 1.$$

Convergence of the continued fraction in a region Ω means convergence of the sequence $\{f_n(z)\}$, the limit of the continued fraction is the limit function $f(z) = \lim_n f_n(z)$. There exists a large body of convergence results for various types of continued fractions, see e.g. [5].

In some situations convergence information of the following kind is available: The sequences of numerators $\{A_n(z)\}$ and denominators $\{B_n(z)\}$ converge to functions $A(z)$ and $B(z)$, or more generally sequences $\{\varphi_n(z)A_n(z)\}$, $\{\varphi_n(z)B_n(z)\}$ converge to functions $C(z)$, $D(z)$, where $\{\varphi_n(z)\}$ is some sequence of functions of a simple form, easily controllable in convergence investigations. In some situations subsequences of $\{\varphi_n(z)A_n(z)\}$, $\{\varphi_n(z)B_n(z)\}$ show similar behavior, as for example in the case of positive PC–fractions and positive Schur fractions. This phenomenon is (non–technically) referred to as <u>separate</u> <u>convergence</u> in the title of this paper. Such separate convergence sometimes makes it possible to draw stronger conclusions about the limit function $f(z)$ than from only "joint" considerations of the approximants $\dfrac{A_n(z)}{B_n(z)}$. Instances where this is of particular importance are the case of positive PC–fractions (with suitable conditions imposed), which are closely connected with orthogonal polynomials on the unit circle, and the case of regular C–fractions (with suitable conditions imposed) and their behavior on the boundary of the convergence region.

In this (mainly expository) paper we discuss conditions for and consequences of the occurrence of the separate convergence phenomenon for some classes of continued fractions.

In all of the following, D denotes the open unit disk $\{z \in C: |z| < 1\}$, $\overline{D}$ denotes the closed unit disk $\{z \in C: |z| \leq 1\}$, and ∂D denotes the unit circle $\{z \in C: |z| = 1\}$.

1. Positive PC–fractions.

A positive PC–fraction is a continued fraction of the form

(1.1)
$$\delta_0 \; - \; \frac{2\delta_0}{1} \; \frac{1}{+ \; \overline{\delta_1}z \; +} \; \frac{(1-|\delta_1|^2)z}{\delta_1} \; \frac{1}{+...+ \; \overline{\delta_n}z \; +} \; \frac{(1-|\delta_n|^2)z}{\delta_n + \; ...,}$$

where $\delta_0 > 0$, $\qquad\qquad |\delta_n| < 1$ for $n = 1,2,...$

The constants δ_n are called the reflection coefficients of the PC–fraction. We denote the numerators and denominators of (1.1) by $P_n(z)$ and $Q_n(z)$. These functions satisfy the recurrence relation

(1.2)
$$\begin{bmatrix} P_{2n} \\ Q_{2n} \end{bmatrix} = \overline{\delta_n}z \begin{bmatrix} P_{2n-1} \\ Q_{2n-1} \end{bmatrix} + \begin{bmatrix} P_{2n-2} \\ Q_{2n-2} \end{bmatrix}, \quad n = 1,2,...$$

(1.3) $\begin{bmatrix} P_{2n+1} \\ Q_{2n+1} \end{bmatrix} = \delta_n \begin{bmatrix} P_{2n} \\ Q_{2n} \end{bmatrix} + (1 - |\delta_n|^2)z \begin{bmatrix} P_{2n-1} \\ Q_{2n-1} \end{bmatrix}$, $n = 1,2,...$

(1.4) $P_0 = \delta_0, \ P_1 = -\delta_0, \ Q_0 = 1, \ Q_1 = 1.$

The even and odd numerators/denominators are connected by the formulas

(1.5) $Q_{2n}(z) = z^n \overline{Q_{2n+1}}(\tfrac{1}{z}), \qquad P_{2n}(z) = -z^n \overline{P_{2n+1}}(\tfrac{1}{z}).$

The denominators $Q_{2n+1}(z)$ and $Q_{2n}(z)$ are Szegö polynomials and reciprocal Szegö polynomials, respectively, for a certain positive linear functional determined by the reflection coefficients. (Szegö polynomials are polynomials orthogonal on the unit circle.)

PC–fractions were introduced in [6]. Their connections with Szegö polynomials were investigated especially in [7,8]. The discussion in this chapter of separate convergence properties is mainly based on [1,2].

We define the constants α_n by

(1.6) $\alpha_0 = \dfrac{1}{\sqrt{\delta_0}}, \quad \alpha_n = \dfrac{1}{\left[\delta_0 \prod\limits_{k=1}^{n} (1-|\delta_k|^2)\right]^{\frac{1}{2}}}$ for $n = 1,2,...$

The determinant formula for (1.1) (See e.g. [5,p.20]) may be written

(1.7) $P_{2n}(z)Q_{2n+1}(z) - P_{2n+1}(z)Q_{2n}(z) = \dfrac{2 \ z^n}{\alpha_n^2}.$

From (1.3) and from (1.2) – (1.3) we obtain the formulas

(1.8) $zQ_{2n-1}(z) = \dfrac{Q_{2n+1}(z) - \delta_n Q_{2n}(z)}{1-|\delta_n|^2}$

(1.9) $Q_{2n-2}(z) = \dfrac{Q_{2n}(z) - \overline{\delta}_n Q_{2n+1}(z)}{1-|\delta_n|^2}.$

Hence

$$(1.10) \quad \begin{cases} \alpha_{m-1}^2[Q_{2m-2}(x)\overline{Q_{2m-2}(y)} - x\bar{y}Q_{2m-1}(x)\overline{Q_{2m-1}(y)}] \\ + \alpha_m^2(1-x\bar{y})Q_{2m+1}(x)\overline{Q_{2m+1}(y)} \\ = \alpha_m^2[Q_{2m}(x)\overline{Q_{2m}(y)} - x\bar{y}Q_{2m+1}(x)\overline{Q_{2m+1}(y)}] \\ \qquad\qquad\qquad\qquad\qquad \text{for } x,y \in C. \end{cases}$$

By adding these formulas for $m = 0,1,2,...n$ (with $Q_{-2} = 0$, $\alpha_{-1} = 0$) we get

$$(1.11) \quad \begin{cases} \alpha_n^2[Q_{2n}(x)\overline{Q_{2n}(y)} - x\bar{y}Q_{2n+1}(x)\overline{Q_{2n+1}(y)}] \\ = (1-x\bar{y}) \sum_{k=0}^{n} \alpha_k^2 \, Q_{2k+1}(x)\overline{Q_{2k+1}(y)}. \end{cases}$$

In the same way we obtain the formula

$$(1.12) \quad \begin{cases} \alpha_n^2[P_{2n}(x)\overline{P_{2n}(y)} - x\bar{y}P_{2n+1}(x)\overline{P_{2n+1}(y)}] \\ = (1-x\bar{y}) \sum_{k=0}^{n} \alpha_k^2 P_{2k+1}(x)\overline{P_{2k+1}(y)}. \end{cases}$$

In particular we get from these formulas by setting $x = y = z$:

$$(1.13) \quad \alpha_n^2(|Q_{2n}(z)|^2 - |z|^2|Q_{2n+1}(z)|^2) = (1-|z|^2) \sum_{k=0}^{n} \alpha_k^2 |Q_{2k+1}(z)|^2$$

$$(1.14) \quad \alpha_n^2(|P_{2n}(z)|^2 - |z|^2|P_{2n+1}(z)|^2) = (1-|z|^2) \sum_{k=0}^{n} \alpha_k^2 |P_{2k+1}(z)|^2.$$

Lemma 1.1. $Q_{2n}(z)$ and $P_{2n}(z)$ have no zeros in $\overline{D}$, and $Q_{2n+1}(z)$ and $P_{2n+1}(z)$ have no zeros in $C - D$.

Proof: It follows from (1.13) that

$$(1.15) \quad \alpha_n^2|Q_{2n}(z)|^2 \geq (1-|z|^2) \, \alpha_0^2|Q_1(z)|^2 = \alpha_0^2(1-|z|^2)$$

and similarly from (1.14) that

$$(1.16) \quad \alpha_n^2|P_{2n}(z)|^2 \geq (1-|z|^2)\alpha_0^2|P_1(z)|^2 = \alpha_0^2(1-|z|^2)\delta_0^2.$$

Since $\alpha_0 \neq 0$, this implies that $Q_{2m}(z) \neq 0$, $P_{2m}(z) \neq 0$ for $z \in D$. It then follows from

(1.5) that $Q_{2m+1}(z) \neq 0$, $P_{2n+1}(z) \neq 0$ for $z \in C{-}\bar{D}$.

Assume that $Q_{2n}(z_0) = 0$, $z_0 \in \partial D$. Then

(1.17) $Q_{2n+1}(z_0) = z_0^n \, \overline{Q_{2n}(z_0)} = 0.$

This contradicts the determinant formula (1.7), hence $Q_{2n}(z) \neq 0$ for $z \in \partial D$. A similar argument shows that $P_{2n}(z) \neq 0$, $Q_{2n+1}(z) \neq 0$, $P_{2n+1}(z) \neq 0$ for $z \in \partial D$.

■

It is known that the even approximants $\dfrac{P_{2n}(z)}{Q_{2n}(z)}$ are Carathéodory functions and converge for $z \in D$ to a Carathéodory function $F(z)$, see [6,7,8]. (A Carathéodory function is a function which is analytic in D and maps D into the half plane $\text{Re} w \geq 0$.) We shall show that under extra conditions on the reflection coefficients δ_n (in addition to $|\delta_n| < 1$, $\delta_0 > 0$) the sequences $\{P_{2n}(z)\}$, $\{P_{2n+1}(z)\}$, $\{Q_{2n}(z)\}$, $\{Q_{2n+1}(z)\}$ converge separately.

Theorem 1.2   Assume that

(1.18) $\displaystyle\sum_{k=1}^{\infty} |\delta_k|^2 < \infty.$

Then the following hold:

A. $\{Q_{2n}(z)\}$ converges for $z \in D$, uniformly on compact subsets, to an analytic function $Q(z)$ without zeros.

B. $\{Q_{2n+1}(z)\}$ converges to zero for $z \in D$.

C. $\{P_{2n}(z)\}$ converges for $z \in D$, uniformly on compact subsets, to an analytic function $P(z)$ without zeros.

D. $\{P_{2n+1}(z)\}$ converges to zero for $z \in D$.

Proof: We first note that because of (1.18), the sequence $\{\alpha_n\}$ converges to a finite limit α, and $\alpha_n \leq \alpha_{n+1} \leq \alpha$.

It follows from Lemma 1.1 that $\dfrac{1}{Q_{2n}(z)}$ is analytic in D. Let $0 < r < 1$. From (1.15) we see that

$$(1.19) \qquad |Q_{2n}(z)| \geq \frac{\alpha_0^{\,2}}{\alpha^2}(1-r^2) \qquad \text{for } |z| \leq r, \ n = 1,2,\ldots$$

From the theory of normal families of analytic functions (see e.g.[3,4]) we conclude that there exists a subsequence $\{Q_{2n(v)}(z)\}$ which converges uniformly on compact subsets of D to a function $Q(z)$ which is analytic and has no zeros.

It follows from (1.13) that

$$(1.20) \qquad \sum_{k=0}^{n(v)} \alpha_k^{\,2}\,|Q_{2k+1}(z)|^2 \leq \frac{\alpha^2|Q_{2n(v)}(z)|^2}{1-|z|^2}.$$

This implies that the monotonic sequence $\left\{ \sum_{k=0}^{n(v)} \alpha_k^{\,2}|Q_{2k+1}(z)|^2 \right\}$ converges for $z \in D$. Then also the sequence $\left\{ \sum_{k=0}^{n} \alpha_k^{\,2}\,|Q_{2k+1}(z)|^2 \right\}$ converges, i.e.

$$(1.21) \qquad \sum_{k=0}^{\infty} \alpha_k^{\,2}|Q_{2k+1}(z)|^2 < \infty \text{ for } z \in D.$$

Since $\alpha_n \xrightarrow[n]{} \alpha \neq 0$, this at once implies that $Q_{2n+1}(z) \xrightarrow[n]{} 0$ for $z \in D$. This proves B.

From (1.11) and the fact that $Q_{2n}(0) = 1$ we obtain by setting $x = z, y = 0$:

$$(1.22) \qquad \alpha_n^{\,2}Q_{2n}(z) = \sum_{k=0}^{n} \alpha_k^{\,2}\, \overline{Q_{2k+1}(0)}\, Q_{2k+1}(z).$$

Let $n > m$. By using the Cauchy–Schwartz inequality we obtain

$$(1.23) \qquad \left\{ \begin{aligned} |\alpha_n^{\,2}Q_{2n}(z) - \alpha_m^{\,2}Q_{2m}(z)|^2 &= \left| \sum_{k=m+1}^{n} \alpha_k^{\,2}\overline{Q_{2k+1}(0)}Q_{2k+1}(z) \right|^2 \\ &\leq \left[\sum_{k=m+1}^{n} \alpha_k^{\,2}|Q_{2k+1}(z)|^2 \right] \cdot \left[\sum_{k+m+1}^{n} \alpha_k^{\,2}|Q_{2k+1}(0)|^2 \right]. \end{aligned} \right.$$

Then (1.21) implies that $\{\alpha_n^{\,2}Q_{2n}(z)\}$ is a Cauchy sequence for $z \in D$. Thus $\{\alpha_n^{\,2}Q_{2n}(z)\}$ converges for $z \in D$, and since the subsequence $\{\alpha_{n(v)}^{\,2} Q_{2n(v)}(z)\}$ converges to $\alpha^2Q(z)$, we may conclude that $\{Q_{2n}(z)\}$ converges to $Q(z)$. It is easily seen that since every subsequence of $\{Q_{2n}(z)\}$ has a subsequence which converges uniformly on compact subsets of D (because of uniform boundedness), the sequence $\{Q_{2n}(z)\}$ must converge uniformly on compact subsets of D. This proves A.

The proofs of C. and D. are completely analogous.

■

We shall next treat a situation where we impose a stronger condition than (1.18). We shall do this by applying an equivalence transformation to (1.1).

<u>Theorem 1.3</u>   <u>The positive PC–fraction (1.1) is equivalent to the continued fraction.</u>

$$(1.24) \quad \delta_0 + \cfrac{1}{-2\delta_0} + \cfrac{1}{-2\cfrac{\delta_1}{\alpha_0}z} + \cfrac{1}{-\frac{1}{2}\alpha_1^2\delta_1\frac{1}{z}} + \ldots + \cfrac{1}{-2\cfrac{\delta_n}{\alpha_{n-1}}z^n} + \cfrac{1}{-\frac{1}{2}\alpha_n^2\delta_n\frac{1}{z^n}}$$

+...

<u>i.e. to the continued fraction</u>

$$(1.25) \quad \delta_0 + \overset{\infty}{\underset{k=1}{K}} \frac{1}{\eth_k} \, ,$$

<u>where</u>

$$(1.26) \quad \eth_1 = -\frac{1}{2\delta_0} \, , \quad \eth_{2n} = -\frac{2}{\alpha_{n-1}^2}\overline{\delta}_n z^n, \quad \eth_{2n+1} = -\frac{1}{2}\alpha_n^2\delta_n\frac{1}{z^n}$$

$$\text{for } n = 1,2,\ldots$$

<u>The numerators $\Pi_n(z)$ and denominators $\Omega_n(z)$ of (1.24) are given by</u>

$$(1.27) \quad \Pi_{2n}(z) = P_{2n}(z) \, , \qquad \Omega_{2n}(z) = Q_{2n}(z)$$

$$(1.28) \quad \Pi_{2n+1}(z) = -\frac{1}{2}\frac{\alpha_n^2}{z^n}P_{2n+1}(z), \qquad \Omega_{2n+1}(z) = -\frac{1}{2}\frac{\alpha_n^2}{z^n}Q_{2n+1}(z).$$

<u>Proof:</u>   Let a_n, b_n be the elements of (1.1). Then

$$(1.29) \quad a_1 = -2\delta_0, \; a_{2n} = 1, \; a_{2n+1} = (1 - |\delta_n|^2)z, \; n = 1,2\ldots$$

(1.30) $b_0 = \delta_0$, $b_1 = 1$, $b_{2n} = \bar{\delta}_n z$, $b_{2n+1} = \delta_n$, $n = 1,2,...$

We define the "equivalence factors" r_n inductively by

(1.31) $r_1 = \dfrac{1}{a_1}$, $r_n = \dfrac{1}{r_{n-1}a_n}$ for $n = 2,3,...$

and set

(1.32) $\tilde{b}_n = r_n b_n$, $n = 1,2,...$

From (1.29), (1.31) we find that

(1.33) $r_1 = -\dfrac{1}{2\delta_0}$, $r_{2n} = \dfrac{1}{r_{2n-1}}$, $r_{2n+1} = \dfrac{r_{2n-1}}{(1-|\delta_n|^2)z}$

and hence

(1.34) $r_{2n} = -2\delta_0(1-|\delta_1|^2)...(1-|\delta_{n-1}|^2)z^{n-1} = -2\dfrac{z^{n-1}}{\alpha_{n-1}^2}$

(1.35) $r_{2n+1} = -\dfrac{1}{2}\dfrac{1}{\delta_0(1-|\delta_1|^2),...(1-|\delta_n|^2)z^n} = -\dfrac{1}{2}\dfrac{\alpha_n^2}{z^n}.$

From this we conclude that the $\tilde{b}_n$ defined by (1.32) may be written in the form (1.26). It follows from [5, p.31–33] that the continued fractions (1.1) and (1.24) are equivalent. We also have the relations

(1.36) $\Pi_{2n}(z) = r_1...r_{2n}P_{2n}(z),$ $\Omega_{2n}(z) = r_1...r_{2n}Q_{2n}(z)$

(1.37) $\Pi_{2n+1}(z) = r_1...r_{2n+1}P_{2n+1}(z),$ $\Omega_{2n+1}(z) = r_1...r_{2n+1}Q_{2n+1}(z)$

so by using (1.33) – (1.35) we obtain the equations (1.27) – (1.28).   ∎

Theorem 1.4.   Assume that

(1.38) $\displaystyle\sum_{k=1}^{\infty} |\delta_k| < \infty.$

Then in addition to A.–D. of Theorem 1.2 the following hold:

E. $\{Q_{2n}(z)\}$ converges uniformly on $\overline{D}$ to a continuous function $Q(z)$.

F. $\{P_{2n}(z)\}$ converges uniformly on $\overline{D}$ to a continuous function $P(z)$.

G.   The function $\dfrac{P(z)}{Q(z)}$, which is a Caratheodory function on D, is

continuous on $\overline{D}$ and has a positive minimum for its real part, i.e.

(1.39) $\displaystyle\min_{z \in \overline{D}} \mathrm{Re}\left[\frac{P(z)}{Q(z)}\right] > 0.$

Proof: Since (1.38) implies (1.18), $\{\alpha_n\}$ tends to a finite limit $\alpha \neq 0$. Consequently there exists a constant K such that

(1.40) $\dfrac{1}{2} |\alpha_n|^2 \leq K, \quad \dfrac{2}{|\alpha_n|^2} \leq K, \; n = 1,2,\ldots$

It then follows from (1.26) that

(1.41) $|\tilde{b}_{2n}| \leq K|\delta_n| \cdot |z|^n, \; |\tilde{b}_{2n+1}| \leq K|\delta_n| \cdot \dfrac{1}{|z|^n}.$

In particular for $z \in \partial D$ we have

(1.42) $|\tilde{b}_{2n}| \leq K|\delta_n|, \; |\tilde{b}_{2n+1}| \leq K|\delta_n|$

and hence

(1.43) $\displaystyle\sum_{n=1}^{\infty} |\tilde{b}_n| < \infty \text{ for } z \in \partial D.$

It follows by the proof of the Stern–Stoltz theorem (see e.g. [5, p.79] that the sequences $\{\Pi_{2n}(z)\}$, $\{\Pi_{2n+1}(z)\}$, $\{\Omega_{2n}(z)\}$, $\{\Omega_{2n+1}(z)\}$ converge uniformly on ∂D. Then by (1.27), also $\{P_{2n}(z)\}$ and $\{Q_{2n}(z)\}$ converge uniformly on ∂D. By the maximum principle ($P_{2n}(z)$

and $Q_{2n}(z)$ are polynomials) we conclude that the sequences $\{P_{2n}(z)\}$ and $\{Q_{2n}(z)\}$ converge uniformly on $\bar{D}$. This proves E and F.

We set

$$(1.44) \qquad a = \prod_{k=1}^{\infty} (1-|\delta_k|), \quad b = \prod_{k=1}^{\infty} (1+|\delta_k|).$$

We observe that a and b are finite numbers different from zero, because of the condition (1.38).

From the recurence relation (1.2) we get

$$(1.45) \qquad \frac{Q_{2n}(z)}{Q_{2n-2}(z)} = 1 + \bar{\delta}_n z \, \frac{Q_{2n-1}(z)}{Q_{2n-2}(z)}.$$

Thus we may write

$$(1.46) \qquad Q_{2n}(z) = \frac{Q_{2n}(z)}{Q_{2n-2}(z)} \cdot \frac{Q_{2n-2}(z)}{Q_{2n-4}(z)} \cdots \frac{Q_2(z)}{Q_0(z)} \cdot Q_0(z) = \prod_{k=1}^{n} \left[1 + \bar{\delta}_k z \, \frac{Q_{2k-1}(z)}{Q_{2k-2}(z)} \right].$$

We note that $\dfrac{Q_{2k-1}(z)}{Q_{2k-2}(z)}$ is analytic for $z \in D$ (cf. Lemma 1.1), and for $z \in \partial D$ we have

$$(1.47) \qquad \left| \frac{Q_{2k-1}(z)}{Q_{2k-2}(z)} \right| = |z|^{k-1} \left| \frac{Q_{2k-2}(z)}{Q_{2k-2}(z)} \right| = 1.$$

Hence by the maximum principle we get

$$(1.48) \qquad \left| \frac{Q_{2k-1}(z)}{Q_{2k-2}(z)} \right| \leq 1 \quad \text{for } z \in \bar{D}.$$

It follows that

$$\prod_{k=1}^{n} (1-|\delta_k|) \leq \prod_{k=1}^{n} \left[1-|\delta_k z \, \frac{Q_{2k-1}(z)}{Q_{2k-2}(z)}| \right] \leq |Q_{2n}(z)|$$

$$\leq \prod_{k=1}^{n} \left[1 + |\delta_k z \, \frac{Q_{2k-1}(z)}{Q_{2k-2}(z)}| \right] \leq \prod_{k+1}^{n} (1 + |\delta_k|).$$

Thus with the definitions (1.44) we get

(1.49) $a \leq |Q_{2n}(z)| \leq b$ for $z \in \bar{D}$.

A completely analogous argument shows that

(1.50) $a \leq |P_{2n}(z)| \leq b$ for $z \in \bar{D}$.

From (1.49) − (1.50) we get

(1.51) $a \leq |Q(z)| \leq b$, $a \leq |P(z)| \leq b$ for $z \in \bar{D}$.

This shows that $P(z)$ and $Q(z)$ have no zeros in $\bar{D}$, so in particular $\dfrac{P(z)}{Q(z)}$ is continuous in $\bar{D}$.

Since $\dfrac{P_{2n}(z)}{Q_{2n}(z)}$ are Carathéodory functions and $\dfrac{P(z)}{Q(z)}$ is analytic for $z \in D$, it follows that $\dfrac{P(z)}{Q(z)}$ is a Carathéodory function, i.e. $\min\limits_{z \in \bar{D}} \dfrac{P(z)}{Q(z)} \geq 0$. We shall now prove the stronger property (1.39).

By using the determinant formula (1.7) on the right hand side of

(1.52) $\mathrm{Re}\begin{bmatrix} \dfrac{P_{2n}(z)}{Q_{2n}(z)} \end{bmatrix} = \dfrac{\frac{1}{2}\{P_{2n}(z)\,\overline{Q_{2n}}(\frac{1}{z}) + \overline{P_{2n}}(\frac{1}{z})Q_{2n}(z)\}}{|Q_{2n}(z)|^2}$

which holds for $z \in \partial D$, we obtain

(1.53) $\mathrm{Re}\begin{bmatrix} \dfrac{P_{2n}(z)}{Q_{2n}(z)} \end{bmatrix} = \dfrac{\frac{1}{2}z^{-n}\{P_{2n}(z)\,Q_{2n+1}(z) - P_{2n+1}(z)Q_{2n+1}(z)\}}{|Q_{2n}(z)|^2} = \dfrac{1}{\alpha_n^2 |Q_{2n}(z)|^2}.$

Consequently by (1.49) − (1.50) we get (letting $n \to \infty$)

(1.54) $\dfrac{1}{\alpha^2 b^2} \leq \mathrm{Re}\begin{bmatrix} \dfrac{P(z)}{Q(z)} \end{bmatrix} \leq \dfrac{1}{\alpha^2 a^2}$ for $z \in \partial D$.

Because of the analyticity, the inequalities are valid for all $z \in \bar{D}$. In particular this implies (1.39).

∎

2. Positive Schur fractions and the Schur algorithm.

A positive Schur fraction is a continued fraction of the form

$$
(2.1) \qquad \gamma_0 + \cfrac{(1-|\gamma_0|^2)z}{\overline{\gamma_0}z} \ \cfrac{1}{+\ \gamma_1 +} \ \cfrac{(1-|\gamma_1|^2)z}{\overline{\gamma_1}z} \ \cfrac{1}{+\dots+\ \gamma_n +} \ \cfrac{(1-|\gamma_n|^2)z}{\overline{\gamma_n}z} \ +\dots\ ,
$$

where $\gamma_0 \in \mathbb{R}$, $|\gamma_n| < 1$ for $n = 1,2,\dots$

We denote the numerators and denominators of (2.1) by $A_n(z)$ and $B_n(z)$. These functions satisfy the recurrence relation

$$
(2.2) \qquad \begin{bmatrix} A_{2n} \\ B_{2n} \end{bmatrix} = \gamma_n \begin{bmatrix} A_{2n-1} \\ B_{2n-1} \end{bmatrix} + \begin{bmatrix} A_{2n-2} \\ B_{2n-2} \end{bmatrix}, \qquad n = 1,2,\dots .
$$

$$
(2.3) \qquad \begin{bmatrix} A_{2n+1} \\ B_{2n+1} \end{bmatrix} = \overline{\gamma_n}z \begin{bmatrix} A_{2n} \\ B_{2n} \end{bmatrix} + (1-|\gamma_n|^2)z \begin{bmatrix} A_{2n-1} \\ B_{2n-1} \end{bmatrix}, \qquad n = 0,1,2\dots,
$$

$$
(2.4) \qquad A_{-1} = 1,\ A_0 = \gamma_0,\ B_{-1} = 0,\ B_0 = 1.
$$

It is well known that the even approximants $\dfrac{A_{2n}(z)}{B_{2n}(z)}$ are Schur functions and converge for $z \in D$ to a Schur function. (A Schur function is a function which is analytic in D and maps D into $\bar{D}$.) Schur functions and their relations to the Schur algorithm which will be introduced presently, are treated in [7]. For the Schur algorithm, see also especially [11].

If a sequence $\{\gamma_n : n = 0,1,2..\}$ is given such that $\gamma_0 \in \mathbb{R}$, $|\gamma_n| < 1$ for $n = 1,2\dots,$ then the Schur algorithm is defined as follows:

Set

$$
(2.5) \qquad t_n(z,w) = \frac{\gamma_n + zw}{1 + \overline{\gamma_n}\,zw}, \qquad n = 0,1,2\dots
$$

$$
(2.6) \qquad T_0(z,w) = t_0(z,w),\ T_n(z,w) = T_{n-1}(z,t_n(z,w)) \quad \text{for } n = 1,2\dots .
$$

The "approximants" of the algorithm may be defined as $T_n(z,0)$, and convergence of the algorithm means convergence of the approximants. In general we may write

$$(2.7) \qquad T_n(z,w) = \frac{zC_n(z)w + D_n(z)}{zE_n(z)w + F_n(z)} \ ,$$

where $C_n(z)$, $D_n(z)$, $E_n(z)$, $F_n(z)$ satisfy the following recurrence relation (see [7,11]):

$$(2.8) \qquad \begin{bmatrix} C_n \\ E_n \end{bmatrix} = z \begin{bmatrix} C_{n-1} \\ E_{n-1} \end{bmatrix} + \overline{\gamma}_n \begin{bmatrix} D_{n-1} \\ F_{n-1} \end{bmatrix}, \qquad n = 1,2,\dots \ .$$

$$(2.9) \qquad \begin{bmatrix} D_n \\ F_n \end{bmatrix} = \gamma_n z \begin{bmatrix} C_{n-1} \\ E_{n-1} \end{bmatrix} + \begin{bmatrix} D_{n-1} \\ F_{n-1} \end{bmatrix}, \qquad n = 1,2,\dots \ ,$$

$$(2.10) \qquad C_0 = 1, \ D_0 = \gamma_0, \ E_0 = \gamma_0, \ F_0 = 1.$$

Then it can be shown that

$$(2.11) \qquad C_n(z) = z^{-1}A_{2n+1}(z), \qquad E_n(z) = z^{-1}B_{2n+1}(z)$$

$$(2.12) \qquad D_n(z) = A_{2n}(z), \qquad F_n(z) = B_{2n}(z).$$

In particular it follows that

$$(2.13) \qquad T_n(z,0) = \frac{D_n(z)}{F_n(z)} = \frac{A_{2n}(z)}{B_{2n}(z)}.$$

Thus the approximants of the Schur algorithm are exactly the even approximants of the positive Schur fraction. Statements about the Schur algorithm are thus the same as statements about the even approximants of the Schur fraction. In particular the approximants of the Schur algorithm converge to a Schur function $f(z)$. Furthermore

$$(2.14) \qquad \gamma_n = f_n(0),$$

where Schur functions $f_n(z)$ are defined inductively by

$$(2.15) \qquad f_0(z) = f(z), \qquad f_{n+1}(z) = \frac{f_n(z) - \gamma_n}{z(1-\overline{\gamma}_n f_n(z))}.$$

On the other hand, every Schur function gives rise to a positive Schur fraction (via the Schur algorithm with f_n defined by (2.14) – (2.15)). The constants γ_n in the Schur fraction are called the Schur parameters.

Let a positive PC–function (1.1) be given. By defining

$$(2.16) \qquad \gamma_0 = \frac{1 - \delta_0}{1 + \delta_0} , \ \gamma_n = \overline{\delta_n} \quad \text{for } n = 1,2\ldots$$

we obtain from the reflection coefficients of the positive PC–fraction the Schur parameters of a positive Schur fraction, and vice versa. The following relationships hold between numerators/denominators of the PC–fraction and the Schur fraction when (2.16) holds $\left[\text{we set } \mu = \dfrac{1}{1 + \delta_0} \right]$:

$$(2.17) \qquad A_{2n} = \mu(Q_{2n} - P_{2n})$$

$$(2.18) \qquad B_{2n} = \mu(Q_{2n} + P_{2n})$$

$$(2.19) \qquad A_{2n+1} = \mu z(Q_{2n+1} - P_{2n+1})$$

$$(2.20) \qquad B_{2n+1} = \mu z\ (Q_{2n+1} + P_{2n+1}).$$

The even approximants $\dfrac{P_{2n}(z)}{Q_{2n}(z)}$ of the positive PC–fraction converge to a Carathéodory function $F(z)$ (see Chapter 1), the even approximants $\dfrac{A_{2n}(z)}{B_{2n}(z)}$ of the positive Schur fraction converges to a Schur function $f(z)$. These functions are related through the formulas

$$(2.21) \qquad f(z) = \frac{1 - F(z)}{1 + F(z)} , \ F(z) = \frac{1 - f(z)}{1 + f(z)} .$$

Similar formulas are valid for the approximants:

$$(2.22) \qquad \frac{A_{2n}}{B_{2n}} = \frac{1 - \dfrac{P_{2n}}{Q_{2n}}}{1 + \dfrac{P_{2n}}{Q_{2n}}} , \ \frac{P_{2n}}{Q_{2n}} = \frac{1 - \dfrac{A_{2n}}{B_{2n}}}{1 + \dfrac{A_{2n}}{B_{2n}}} .$$

For details on these relationships, see especially [7].

We may now use the results of Chapter 1 to obtain stronger convergence results and results on the limiting functions when stronger conditions are imposed on the Schur parameters γ_n.

Theorem 2.1.   Assume that

$$(2.23) \qquad \sum_{k=1}^{\infty} |\gamma_k|^2 < \infty.$$

Then the following hold:

A. $\{B_{2n}(z)\}$ converges on D, uniformly on compact subsets, to an analytic function B(z) without zeros.

B. $\{B_{2n+1}(z)\}$ converges to zero for $z \in D$.

C. $\{A_{2n}(z)\}$ converges on D, uniformly on compact subsets, to an analytic function A(z).

D. $\{A_{2n+1}(z)\}$ converges to zero for $z \in D$.

Proof: The convergence results follow immediately from (2.17) – (2.20) and Theorem 1.2. (for the PC–fraction whose reflection coefficients are defined by (2.16)). We may write $B_{2n}(z)$ as

$$(2.24) \qquad B_{2n}(z) = \mu Q_{2n}(z) \left[1 + \frac{P_{2n}(z)}{Q_{2n}(z)} \right].$$

Since Re $\left[\dfrac{P_{2n}(z)}{Q_{2n}(z)} \right] \geq 0$ and $Q_{2n}(z) \neq 0$, $B_{2n}(z)$ can have no zeros in D. It then follows e.g. by Hurwitz' theorem (see [3,4]) that B(z) has no zeros in D.   ∎

The results of Theorem 2.1 can essentially be found in [9, p.111], there obtained by a different argument.

Theorem 2.2.   Assume that

$$(2.25) \qquad \sum_{k=1}^{\infty} |\gamma_k| < \infty.$$

Then in addition to A. – D. of Theorem 2.1 the following hold:

E. $\{B_{2n}(z)\}$ converges uniformly on $\overline{D}$ to a continuous function $B(z)$.

F. $\{A_{2n}(z)\}$ converges uniformly on $\overline{D}$ to a continuous function $A(z)$.

G.   The function $\dfrac{A(z)}{B(z)}$, which is a Schur function on D, is continuous on $\overline{D}$ and has a maximum for its absolute value less than one, i.e.

$$(2.26) \qquad \max_{z \in \overline{D}} |f(z)| < 1.$$

Proof: The convergence results follow immediately from (2.17) – (2.20) and Theorem 1.4 (again for the PC–fraction whose reflection coefficients are defined by (2.16)).

Because of Theorem 1.4 G and (2.21) where $F(z) = \dfrac{P(z)}{Q(z)}$, the function $f(z) = \dfrac{A(z)}{B(z)}$ is continuous on $\overline{D}$. Because of (1.39) the continuous function $F(z) = \dfrac{P(z)}{Q(z)}$ maps $\overline{D}$ onto a compact subset K of the open right half plane. The mapping $F \longrightarrow \dfrac{1 - F}{1 + F}$ maps K onto a compact subset K' of D. This means that the function $f(z) = \dfrac{A(z)}{B(z)} = \dfrac{1 - \dfrac{P(z)}{Q(z)}}{1 + \dfrac{P(z)}{Q(z)}}$ maps $\overline{D}$ onto the compact subset K' of D, which implies (2.26).

∎

By arguments related to those used to obtain (1.54) it can be shown that the compact set K in the proof of Theorem 2.2 is contained in the disk with diameter $[\dfrac{1}{L}, L]$ where $L = \dfrac{b}{a}$. By elementary properties of the linear fractional transformation $F \longrightarrow \dfrac{1 - F}{1 + F}$ it then follows that the compact set K' is contained in the disk with center at the origin and radius $\dfrac{L - 1}{L + 1}$. (See [1].)

The results of Theorem 2.2 G were proved by Schur in [11]. They were obtained by arguments different from those we have presented, which are mainly based on [1,2]. See also [9].

We shall now consider a result proved by Runckel in [9], where the conditions on the Schur parameters are stronger than (2.23), weaker than (2.25). We only give a sketch of the proof.

We shall use the following notation. For any natural number p we set

(2.27) $\quad R_p = \{z \in C: z^p = 1\}.$

For a given sequence $\{\gamma_n\}$ of Schur parameters we set

(2.28) $\quad \beta_n = \left[\prod_{k=0}^{n} (1 - |\gamma_k|^2) \right]^{\frac{1}{2}}.$

It follows from (1.5) together with (2.17) – (2.20) that the even/odd numerators/denominators of the Schur fraction are connected by the formulas

(2.29) $\quad B_{2n}(z) = z^{n+1}\overline{A_{2n+1}}(\frac{1}{z}), \qquad A_{2n}(z) = z^{n+1}\overline{B_{2n+1}}(\frac{1}{z}).$

The determinant formula for the Schur fraction may then be written

(2.30) $\quad B_{2n}(z)\overline{B_{2n}}(\frac{1}{z}) - A_{2n}(z)\overline{A_{2n}}(\frac{1}{z}) = \beta_n^{2} \qquad$ for $x \in \partial D$,

and hence

(2.31) $\quad |B_{2n}(z)|^2 - |A_{2n}(z)|^2 = \beta_n^{2} \qquad$ for $z \in \partial D$.

We recall that in addition to the determinant formula also the following general formula is valid (see e.g. [5, p.20]):

(2.32) $\quad - A_n B_{n+2} + A_{n+2}B_n = b_{n+2} \cdot \pi_n,$

where b_{n+2} is the denominator element, π_n is the right hand side in the determinant formula $A_{n+1}B_n - A_n B_{n+1} = \pi_n.$

For Schur fractions this equality reads

(2.33) $\quad A_{2n}B_{2n+2} - A_{2n+2}B_{2n} = \gamma_{n+1}\beta_n^{2}z^{n+1}.$

<u>Theorem 2.3.</u>   <u>Assume that</u>

(2.34) $\quad \sum_{n=1}^{\infty} |\gamma_n|^2 < \infty$ <u>and</u> $\sum_{n=1}^{\infty} |\gamma_{n+p} - \gamma_n| < \infty$ <u>for some</u> $p \in \mathbb{N}.$

Then the even approximants $\dfrac{A_{2n}(z)}{D_{2n}(z)}$ converge uniformly on compact subsets of $\overline{D} - R_p$ to a function $f(z) = \dfrac{A(z)}{D(z)}$ which is (analytic on D and) continuous on $\overline{D} - R_p$, and such that

$$(2.35) \qquad |f(z)| < 1 \quad \text{for } z \in \overline{D} - R_p.$$

Proof.   (Sketch). By using the formula (2.33) we find that we may write

$$(2.36) \quad
\begin{cases}
(1 - z^p)\,\dfrac{A_{2N}(z)}{B_{2N}(z)} = (1 - z^p)\gamma_0 + \displaystyle\sum_{n=0}^{p-1} \dfrac{\gamma_{n+1}\beta_n^2 z^{n+1}}{B_{2n}(z)B_{2n+2}(z)} \\[3ex]
\quad - \displaystyle\sum_{n=N-p}^{N-1} \dfrac{\gamma_{n+1}\beta_n^2 z^{n+p-1}}{B_{2n}(z)B_{2n+2}(z)} \\[3ex]
\quad + z^{p+1} \displaystyle\sum_{n=0}^{N-p-1} \left[\dfrac{\gamma_{n+p+1}\beta_{n+p}^2}{B_{2n+2p}(z)B_{2n+2p+2}(z)} \right. \\[3ex]
\qquad\qquad\qquad \left. - \dfrac{\gamma_{n+1}\beta_n^2}{B_{2n}(z)B_{2n+2}(z)} \right] z^n.
\end{cases}$$

The first sum on the right is a fixed finite sum. The second sum consists of p terms, which can be seen to tend to zero uniformly for $z \in \overline{D}$ (by using (2.31) and the relations $\beta_n^2 \to \beta^2 \neq 0$, $\gamma_{n+1} \to 0$). By using the condition $\displaystyle\sum_{n=1}^{\infty} |\gamma_{n+p} - \gamma_n| < \infty$ it can be shown that the third sum converges uniformly for $z \in \overline{D}$. (See [9, p.105].)

It thus follows that $\left\{ (1-z^p)\dfrac{A_{2N}(z)}{B_{2N}(z)} \right\}$ converges uniformly on $\overline{D}$. Consequently $\left\{ \dfrac{A_{2N}(z)}{D_{2N}(z)} \right\}$ converges to a continuous function for $z \in \overline{D} - R_p$.

It can be shown that there exists for every compact subset K of $\overline{D} - R_p$ a uniform bound M_K for $B_{2n}(z)$:

$$(2.37) \qquad |B_{2n}(z)| \leq M_K \quad \text{for } z \in K.$$

(For proof, see [9 p. 99–104].) We may write (2.31) in the form

$$(2.38) \qquad 1 - \left|\frac{A_{2n}(z)}{B_{2n}(z)}\right|^2 = \frac{\beta_n^{\,2}}{|B_{2n}(z)|^2} \qquad \text{for } z \in \partial D$$

and from this by using (2.37) we get

$$(2.39) \qquad \left|\frac{A_{2n}(z)}{B_{2n}(z)}\right|^2 | \leq 1 - \frac{\beta^2}{M_K^{\,2}} \qquad \text{for } z \in K.$$

Consequently also

$$(2.40) \qquad |f(z)|^2 \leq 1 - \frac{\beta^2}{M_K^{\,2}} \qquad \text{for } z \in K.$$

In particular $|f(z)| \leq 1 \quad$ for $z \in \overline{D} - R_p$.

$\blacksquare$

3. General T–fractions.

A general T–fraction is a continued fraction of the form

$$(3.1) \qquad \frac{F_1 z}{1 + G_1 z} + \frac{F_2 z}{1 + G_2 z} + \ldots + \frac{F_n z}{1 + G_n z} + \ldots, \quad F_n \neq 0 \text{ for } n = 1,2,\ldots .$$

In [10] Runckel considered implicitely general T–fractions of the form

$$(3.2) \qquad \frac{(b_1 - 1)z}{1 + z} + \frac{(b_2 - 1)z}{1 + z} + \ldots + \frac{(b_n - 1)z}{1 + z} + \ldots, \quad b_n \neq 1 \text{ for } n = 1,2,\ldots .$$

that is: general T–fractions with $G_n = 1$ for all n. We shall call these fractions general T–fractions of Runckel's type. We shall denote the approximants by $\dfrac{R_n(z)}{S_n(z)}$. The following recurrence relation is then satisfied:

$$(3.3) \qquad \begin{bmatrix} R_n \\ S_n \end{bmatrix} = (1 + z) \begin{bmatrix} R_{n-1} \\ S_{n-1} \end{bmatrix} + (b_n - 1)z \begin{bmatrix} R_{n-2} \\ S_{n-2} \end{bmatrix}, \quad n = 1,2,\ldots ,$$

$$(3.4) \qquad R_{-1} = 1, \quad R_0 = 0, \quad S_{-1} = 0, \quad S_0 = 1.$$

(Runckel worked with functions $C_n(z)$, $D_n(z)$ satisfying a recurrence relation of the form (3.3), but with initial values $C_0 = 0$, $D_0 = 2 - z$, $C_1 = 1 - z$, $D_1 = 1 - z^2$. For questions of convergence this makes no difference, except for $z = -1$.)

We state without proof a result that will be referred to in Chapter 4. The result is contained in [10, Lemmas 2,3].

Theorem 3.1.   Assume that

$$(3.5) \qquad \sum_{n=1}^{\infty} |b_n| < \infty$$

in the continued fraction (3.2). Then the sequences $\{R_n(z)\}$, $\{S_n(z)\}$ converge, uniformly on compact subsets of D, to functions $R(z)$, $S(z)$ which are (analytic on D and) continuous on $\overline{D}$, except possibly for $z = 1$.

If the stronger condition

$$(3.6) \qquad \sum_{n=1}^{\infty} n |b_n| < \infty$$

is satisfied, then the functions $R(z)$, $S(z)$ are continuous on $\overline{D}$.

In connection with these comments on general T–fractions we also mention the following. A T–fraction is a general T–fraction where $F_n = 1$ for all n, i.e. a continued fraction that may be written in the form

$$(3.7) \qquad \frac{z}{1 + d_1 z} + \frac{z}{1 + d_2 z} + \ldots + \frac{z}{1 + d_n z} + \ldots \, .$$

We denote by $\dfrac{U_n(z)}{V_n(z)}$ the approximants, where $U_n(z)$, $U_n(z)$ are defined by a recurrence relation similar to (3.3) – (3.4).

In [14] Waadeland studied convergence of T–fractions corresponding to certain functions which are analytic at the origin. From this paper the following result can easily be obtained (see[14, p. 10–13]):

Theorem 3.2.   Assume that

(3.8) $|d_n + 1| \le c\lambda^n$ for some $\lambda \in (0,1)$

in the T–fraction (3.7). Then the sequences $\{U_n(z)\},\{V_n(z)\}$ converge, uniformly on compact subsets of D, to analytic functions $U(z)$, $V(z)$.

4. Regular C–fractions and g–fractions.

We consider regular C–fractions written in the form

(4.1) $$\frac{a_0}{1} - \frac{a_1 z}{1} - \frac{a_2 z}{1} - \ldots - \frac{a_n z}{1} - \ldots$$

We write $\dfrac{\Gamma_n(z)}{\Delta_n(z)}$ for the approximants. The following recurrence relation is satisfied:

(4.2) $\begin{bmatrix}\Gamma_n \\ \Delta_n\end{bmatrix} = \begin{bmatrix}\Gamma_{n-1} \\ \Delta_{n-1}\end{bmatrix} - a_{n-1}z\begin{bmatrix}\Gamma_{n-2} \\ \Delta_{n-2}\end{bmatrix}$, $n = 2,3,\ldots$,

(4.3) $\Gamma_0 = 0, \Gamma_1 = a_0, \Delta_0 = 1, \Delta_1 = 1.$

We assume that the continued fraction is limit periodic, i.e. that $\lim_n a_n = a$ exists. We only consider the more difficult case $a \ne 0$, and shall without loss of generality set $a = \frac{1}{4}$, i.e.

(4.4) $\lim_n a_n = \frac{1}{4}.$

We define the critical line L by

(4.5) $L = \{z \in C: z \in \mathbb{R}, z \ge 1\}.$

When in the following we talk about the closed cut plane $\overline{C-L}$ and the boundary $\partial(C-L) = (\overline{C-L}) - (C-L)$, it is understood that to each point on $L - \{1\}$ there belong two points, one on the "upper side of the cut", one on the "lover side of the cut".

It is well known that the approximants $\dfrac{\Gamma_n(z)}{\Delta_n(z)}$ converge to a function meromorphic in C–L. (See e.g. [5, p.132].) We shall show that under extra conditions on the elements a_n, stronger statements about the convergence and the limit functions can be obtained.

In the following we set

(4.6) $b_n = 1 - 4a_n$

and note that

(4.7) $\lim_n b_n = 0$.

We make the following transformation of the variable z into a variable w:

(4.8)
$$z = \frac{4w}{(1 + w)^2}$$

$$w = \frac{2}{z} - 1 - \frac{2}{z} \sqrt{1 - z}, \qquad \sqrt{1 - z} > 0 \text{ for } z < 1.$$

This transformation maps the region C–L in the z–plane biuniquely onto the open unit disk $D = \{w : |w| < 1\}$ in the w–plane. The boundary $\partial(C–L)$ (in the meaning indicated above) is mapped biuniquely onto the unit circle $\partial D = \{w : |w| = 1\}$.

To each function f(z) on C–L corresponds a function F(w) on D, defined by

(4.9) $F(w) = f(z)$

(z and w being related as in (4.8)), and vice versa. Boundary behavior of F(w) at ∂D can be directly translated into boundary behavior of f(z) at the boundary $\partial(C–L)$.

By the substitutions (4.6) and (4.8) the continued fraction (4.1) is transformed into the continued fraction

(4.10) $\dfrac{a_0}{1} + \dfrac{(b_1-1)w(1+w)^{-2}}{1} + ... + \dfrac{(b_n-1)w(1+w)^{-2}}{1} + ...$

Let x_n, y_n be the elements of (4.10), i.e.

$$(4.11) \quad x_1 = a_0, \quad x_n = \frac{(b_{n-1}-1)w}{(1+w)^2} \quad \text{for } n = 2,3,...$$
$$y_n = 1 \quad \text{for } n = 1,2... \; .$$

Let the numerators and denominators of the approximants be denoted by $X_n(w)$, $Y_n(w)$. Then

$$(4.12) \quad X_n(w) = \Gamma_n(z), \; Y_n(w) = \Delta_n(z).$$

Theorem 4.1. The continued fraction (4.10) is equivalent to the continued fraction

$$(4.13) \quad \frac{a_0(1+w)}{1+w} \quad + \quad \frac{(b_1-1)w}{1+w} \quad +...+ \quad \frac{(b_n-1)w}{1+w} \quad +... \; .$$

Its numerators $\tilde{X}_n(w)$ and denominators $\tilde{Y}_n(w)$ are given by

$$(4.14) \quad \tilde{X}_n(w) = (1 + w)^n X_n(w), \; \tilde{Y}_n(w) = (1 + w)^n Y_n(w).$$

Proof: Let $\tilde{x}_n$, $\tilde{y}_n$ be the elements of (4.13), i.e.

$$(4.15) \quad \tilde{x}_1 = a_0(1 + w), \; \tilde{x}_n = \frac{(b_{n-1}-1)w}{1 + w} \quad \text{for } n = 2,3,... .$$
$$\tilde{y}_n = 1 + w \quad \text{for } n = 1,2,... \; .$$

Set

$$(4.16) \quad r_n = (1 + w) \quad \text{for } n = 1,2... \; .$$

We observe that

$$(4.17) \quad \tilde{x}_1 = r_1 x_1, \; \tilde{x}_n = r_n r_{n-1} x_n \quad \text{for } n = 2,3,...$$
$$\tilde{y}_n = r_n y_n \quad \text{for } n = 1,2,... \; .$$

It follows (see e.g. [5, 31–33]) that the continued fractions (4.10) and (4.13) are equivalent. The formulas (4.14) follow from the fact that

$$(4.18) \quad \tilde{X}_n = r_1 ... r_n X_n, \; \tilde{Y}_n = r_1 ... r_n Y_n.$$

Theorem 4.2.   Assume that

(4.19) $\displaystyle\sum_{n=1}^{\infty} |b_n| < \infty$

in the continued fraction (4.10). (Recall that $\Gamma_n(z)$, $\Delta_n(z)$ denote the numerators and denominators of the approximants, and that w is defined in terms of z by (4.8).) Then $\{(1+w)^n \Gamma_n(z)\}$ and $\{(1+w)^n \Delta_n(z)\}$ converge, uniformly on compact subsets of C–L, to functions $\Gamma_\infty(z)$, $\Delta_\infty(z)$ which are (analytic on C–L and) continuous on $\overline{\text{C–L}}$, except possibly at the point $z = 1$.

If the stronger condition

(4.20) $\displaystyle\sum_{n=1}^{\infty} n|b_n| < \infty$

is assumed, then the functions $\Gamma_\infty(z)$, $\Delta_\infty(z)$ are continuous on C–L.

Proof: Except for the first term, (4.13) is a general T–fraction of Runckel's type. Therefore the results in Theorem 3.1 hold for this continued fraction. The results of Theorem 4.2 follow by applying Theorem 4.1, the mapping properties of the transformation (4.8) (including the fact that $z = 1$ corresponds to $w = 1$) being taken into account.

∎

For results on separate convergence for the continued fraction (4.1), including the first part of Theorem 4.2, see also [12,13].

We now consider the situation that the sequence $\{a_n\}$ in (4.1) is a chain sequence, i.e. that

(4.21) $\begin{cases} a_n = (1 - g_{n-1})g_n, & 0 < g_n < 1 \quad \text{for } n = 1,2,... \\ a_0 = g_0 > 0. \end{cases}$

The regular C–fraction (4.1) is in this case a g–fraction, and has the form

(4.22) $\dfrac{g_0}{1} - \dfrac{(1-g_0)g_1 z}{1} - ... - \dfrac{(1-g_{n-1})g_n z}{1} - ...$,

where $g_0 > 0,$ $0 < g_n < 1$ for $n = 1,2,...$.

It is well known (see e.g. [10]) that

$$(4.23) \qquad \lim_{n} a_n = \frac{1}{4} \Longleftrightarrow \lim_{n} g_n = \frac{1}{2}.$$

Theorem 4.3. Let the g–fraction (4.22) be given, and let a_n be defined by (4.21), b_n by (4.6). Then the following two conditions are equivalent:

$$(4.24) \qquad \sum_{n=1}^{\infty} |b_n| < \infty$$

$$(4.25) \qquad \sum_{n=1}^{\infty} |g_n - \tfrac{1}{2}|^2 < \infty \quad \underline{and} \quad \sum_{n=1}^{\infty} |g_{n+1} - g_n| < \infty.$$

Proof: Straightforward calculation shows that the following identities hold:

$$(4.26) \qquad a_n - \frac{1}{4} = g_n(g_n - g_{n-1}) - (g_n - \tfrac{1}{2})^2$$

$$(4.27) \qquad a_n - \frac{1}{4} = \frac{1}{2}(g_n - g_{n-1}) - (g_n - \tfrac{1}{2})(g_{n-1} - \tfrac{1}{2})$$

$$(4.28) \qquad (a_n - \tfrac{1}{4})^2 = \frac{1}{4}(g_n - \tfrac{1}{2})^2 - g_n(g_n - \tfrac{1}{2})(g_{n-1} - \tfrac{1}{2}) + g_n^2(g_{n-1} - \tfrac{1}{2})^2.$$

We first show that (4.25) implies (4.24). It follows from (4.26) that

$$(4.29) \qquad \sum_{n=1}^{\infty} |a_n - \tfrac{1}{4}| < \infty$$

when (4.25) holds. This is the same as (4.24).

We next show that (4.24) implies (4.25). From (4.28) we conclude that

$$(4.30) \qquad -(g_n - \tfrac{1}{2})(g_{n-1} - \tfrac{1}{2}) \leq M |a_n - \tfrac{1}{4}|^2$$

for some constant M, since the sequence $\{g_n\}$ has a positive lower bound. (Note that (4.24) implies $a_n \longrightarrow \frac{1}{4}$, hence $g_n \longrightarrow \frac{1}{2}$ by (4.23).) It follows that

$$(4.31) \qquad - \sum {}'(g_n - \tfrac{1}{2})(g_{n-1} - \tfrac{1}{2}) < \infty,$$

where only non–positive terms are included in the sum.

From (4.27) we see that

$$(4.32) \qquad \sum_{n=1}^{\infty} (a_n - \tfrac{1}{4}) = \tfrac{1}{2}(\tfrac{1}{2} - g_0) - \sum_{n=1}^{\infty} (g_n - \tfrac{1}{2})(g_{n-1} - \tfrac{1}{2}).$$

It then follows from (4.24) that the series $\sum_{n=1}^{\infty} (g_n - \tfrac{1}{2})(g_{n-1} - \tfrac{1}{2})$ converges. Taking into account (4.31) we may conclude that

$$(4.33) \qquad \sum {}''(g_n - \tfrac{1}{2})(g_{n-1} - \tfrac{1}{2}) < \infty,$$

where only positive terms are included in the sum. Hence

$$(4.34) \qquad \sum_{n=1}^{\infty} |g_n - \tfrac{1}{2}| \cdot |g_{n-1} - \tfrac{1}{2}| < \infty.$$

From (4.27) now follows that $\sum_{n=1}^{\infty} |g_n - g_{n-1}| < \infty$, and from (4.26) follows (since $|g_n| <$

1) that $\sum_{n=1}^{\infty} |g_n - \tfrac{1}{2}|^2 < \infty$.

■

From Theorem 4.2 we now immediately obtain the following result for g–fractions (cf. [10]).

Theorem 4.4.   Let the g–fraction (4.22) be given, and assume that (4.25) is satisfied. Let $\Gamma_n(z)$, $\Delta_n(z)$ denote the numerators and denominators of the approximants, and let w be defined in terms of z by (4.8). Then $(1 + w)^n \Gamma_n(z)$ and $(1 + w)^n \Delta_n(z)$ converge on C–L, uniformly on compact subsets, to functions $\Gamma_\infty(z)$, $\Delta_\infty(z)$ which are (analytic in C–L and) continuous in $\overline{C-L}$, except possibly at z = 1.

As a special case of this theorem we have the following (see [9,15]):

Corollary 4.5.   Let the g–fraction (4.22) be given and assume that the condition

$$(4.35) \qquad \sum_{n=1}^{\infty} |g_n - \tfrac{1}{2}| < \infty$$

is satisfied. Then the conclusions of Theorem 4.4 hold.

Proof: Condition (4.35) implies (4.25).   ∎

Analyticity properties and continuity properties of the limit function $\dfrac{\Gamma_\infty(z)}{\Delta_\infty(z)}$ in
Theorem 4.4 and Corollary 4.5 follow immediately. These results are closely related to a
theorem of Wall [9,15]. Wall's approach is completely different from the one presented
here, and the results are slightly different. The method consists in transforming results for
the Schur algorithm in Theorem 2.2 into results on the g–fraction, the condition (2.25)
being transformed into the condition (4.35). Runckel adapted Wall's approach in [9] to
obtain results essentially containing the results on $\dfrac{\Gamma_\infty(z)}{\Delta_\infty(z)}$, obtained from Theorem 4.4. In
that paper, the results in Theorem 2.3 are transformed into results on g–fractions, with the
condition (2.34) translated into the condition $\sum\limits_{n=1}^{\infty} |g_n - \tfrac{1}{2}|^2 < \infty$ and $\sum\limits_{n=1}^{\infty} |g_{n+p} - g_n| < \infty$.
So in that respect the situation is somewhat more general than in Theorem 4.4. However,
these arguments of Wall and Runckel do not give results about separate convergence of
numerators and denominators and regularity results for their limit functions, but only
results about the approximants and their limits. Our proof of Theorem 4.2 and its
consequence Theorem 4.4 is an adaption of Runckel's treatment in [10].

References.

1. Geronimus, Ya.L: Polynomials orthogonal on a circle and their applications, Amer.Math.Soc., Translation Number 104, Providence (1954).

2. Geronimus, Ya.L.: Orthogonal Polynomials, Consultants Bureau, New York (1961).

3. Henrici, P.: Applied and Computational Complex Analysis, vol I, Wiley, New York (1974).

4. Hille, E.: Analytic Function Theory, vol II, Ginn, New York (1962).

5. Jones, W.B. and Thron, W.J.: Continued Fractions: Analytic Theory and Applications, Encyclopedia of Mathematics and its Applications, 11, Addison–Wesley, Reading (1980), distributed now by Cambridge University Press.

6. Jones, W.B, Njåstad, O. and Thron, W.J.: Continued fractions associated with the trigonometric and other strong moment problems, Constructive Approximation 2 (1986) 197–211.

7. Jones, W.B., Njåstad, O. and Thron, W.J.: Schur fractions, Perron–Carathéodory fractions and Szegö polynomials, a Survey, Analytic Theory of Continued fractions II, W.T. Thron (ed.), Springer Lecture Notes in Mathematics 1199, New York – Berlin (1986) 127–158.

8. Jones, W.B., Njåstad, O. and Thron, W.J.: Moment theory, orthogonal polynomials, quadrature and continued fractions associated with the unit circle, Bull. Lond. Math. Soc., to appear.

9. Runckel, H.: Bounded analytic functions in the unit disk and the behavior of certain analytic continued fractions near the singular line, J. Reine Angew. Math. 281 (1976) 97–125.

10. Runckel, H.: Continurity on the boundary and analytic continuation of continued fractions, Math. Zeitschrift 148 (1976) 189–205.

11. Schur, I.: Über Potenzreihen, die im Innern des Einheitskreises beschränkt sind, J. Reine Angew. Math. 147 (1916) 205–232, 148(1917) 122–145.

12. Sleschinsky, J.W.: Über die Convergenz der Kettenbrüche, Odessa. Ges. VIII (1888) 97–127.

13. Sleschinsky, J.W.: Beweis der Existenz einiger Grentzen, Oddessa. Ges. VIII (1888) 127–137.

14. Waadeland, H.: A convergence property of certain T–fraction expansions, Kgl. Norske Videnskabers Selskab, Skrifter No. 9 (1966) 1–22.

15. Wall, H.S.: The behavior of certain Stieltjes continued fractions near the singular line, Bull. Amer. Math. Soc. 51 (1942) 427–431.

SOME REMARKS ON NEARNESS PROBLEMS FOR
CONTINUED FRACTION EXPANSIONS

Olav Njåstad
The University of Trondheim NTH
Division of Mathematical Sciences
N–7034 Trondheim–NTH
Norway

Haakon Waadeland
Department of Mathematics and Statistics
The University of Trondheim AVH
N–7055 Dragvoll
Norway

Abstract. In some cases "nearness" between functions implies "nearness" between their possible continued fraction expansions of some type. This of course depends strongly upon the way "nearness" between functions is defined, but also upon the choice of function to be "near". The present note is a brief, informal and example-based discussion of this matter.

In representing analytic functions by continued fractions the following type of question has been raised on several occations: Will "nearness" between functions imply "nearness" between the possible continued fraction expansions? In the different answers the word "nearness" has been replaced by well defined properties which may be thought of as being "nearness properties". Examples 1–3 are meant to serve as illustrations of this.

Example 1. It is a simple and well known fact that the pair $(z, -1)$ of functions may be represented by the general T-fraction

$$\cfrac{z}{1-z} + \cfrac{z}{1-z} + \cdots + \cfrac{z}{1-z} + \cdots,$$

meaning that the continued fraction corresponds to

$$z + 0z^2 + 0z^3 + \cdots$$

at 0 and to

$$-1 + 0z^{-1} + 0z^{-2} + \cdots$$

at ∞. For definition of correspondence in this case we refer to [3, Sec. 5.1]. In [5] it is proved, that if

$$L_o(z) = c_1 z + c_2 z^2 + \cdots, c_1 \neq 0,$$

is holomorphic and sufficiently close to $c_1 z$ in a large enough neighborhood of 0, and

$$L_\infty(z) = -c_o - c_{-1} z^{-1} - \cdots$$

is holomorphic and sufficiently close to -1 in a large enough neighborhood of ∞, then

$(L_o(z), L_\infty(z))$ may be represented in a similar way by a general T-fraction

$$\mathop{\text{K}}_{n=1}^{\infty} \frac{F_n z}{1 + G_n z},$$

where

$$\lim_{n \to \infty} F_n = F \neq 0$$

exists, and where

$$\lim_{n \to \infty} G_n = -F.$$

(For a precise statement see [5, Sec. 2].) Observe that for any a (for instance for $a = c_1$ or $a = F$) the general T-fraction $\mathop{\text{K}}_{n=1}^{\infty}(az/(1 - az))$ corresponds to $(az, -1)$. Observe furthermore, that the value of c_1 is inessential in the following sense: Nearness of $L_o(z)$ to $c_1 z$ is essentially the same as nearness of $K \cdot L_o(z)$ to $K \cdot c_1 z$, for instance: Nearness of $(F/c_1) \cdot L_o(z)$ to Fz.

Example 2. It is well known, that the function

$$\frac{1}{2}(\sqrt{1 + 4w} - 1), \quad \Re\sqrt{1 + 4w} > 0,$$

may be represented by the regular C-fraction

$$\mathop{\text{K}}_{n=1}^{\infty} \left(\frac{w}{1}\right).$$

In [1] it is proved that if $g(z)$ is holomorphic and sufficiently small in a sufficiently large neighborhood of 0, and $g(0) = 0$, then

$$\frac{\sqrt{1 + 4w} - 1}{2} \left(1 + g\left(\frac{2w}{1 + 2w + \sqrt{1 + 4w}}\right)\right)$$

has a regular C-fraction expansion $\text{K}(a_n w/1)$, where $a_n \to 1$ when $n \to \infty$. (The function

$$z = \frac{2w}{1 + 2w + \sqrt{1 + 4w}},$$

being the inverse of the Koebe function, maps the complement of the ray $(-\infty, -\frac{1}{4}]$ one to one onto the open unit disk U.)

Example 3. Here we deal with the PC-fractions

$$\frac{1}{2} - \frac{1}{1 + \beta_2 z} + \frac{(1 - \beta_2 \beta_3)z}{\beta_3} + \frac{1}{\beta_4 z} + \frac{(1 - \beta_4 \beta_5)z}{\beta_5} + \cdots$$

The relevant type of correspondence is to a pair of power series of the form

$$L_o(z) = \frac{1}{2} + c_1 z + c_2 z^2 + \cdots,$$

$$L_\infty(z) = -\frac{1}{2} - c_{-1}z^{-1} - c_{-2}z^{-2} - \cdots,$$

such that the sequence of even approximants corresponds at 0 to $L_o(z)$ and the sequence of odd approximants to $L_\infty(z)$ at ∞. For a precise definition as well as conditions for correspondence we refer to [2]. We recall that if in addition all $\beta_{2n} \neq 0$ then the even approximants also correspond to $L_\infty(z)$ (at ∞), and if all $\beta_{2n+1} \neq 0$ then the odd approximants correspond to $L_o(z)$ (at 0)[2, Thm. 4.1.]. One very special example is obtained when $L_o(z)$ and $L_\infty(z)$ reduce to merely $1/2$ and $-1/2$, in which case it is easily seen that we have the PC-fraction representation

$$\frac{1}{2} - \frac{1}{1} + \frac{1}{0z} + \frac{z}{0} + \frac{1}{0z} + \frac{z}{0} + \cdots.$$

In [4] it is proved, that if $(L_o(z), L_\infty(z))$ is sufficiently close to $(1/2, -1/2)$, meaning that $L_o(z) - 1/2$ is sufficiently small in a large enough neighborhood of 0 , and $L_\infty(z) + 1/2$ is sufficiently small in a large enough neighborhood of ∞, then $(L_o(z), L_\infty(z))$ has a PC-fraction representation where $\beta_n \to 0$ when $n \to \infty$.

In the examples shown above, as well as in other results, nearness of functions or pairs of functions always meant nearness to a very specific and simple reference function (or pair of functions). A rather natural guess would be to assume that the role of such a reference function or reference pair merely is to simplify argument or computation, and that similar results will hold with more general reference functions and the same type of nearness condition. As will be seen in the present note, however, the role of the reference functions is more essential than just that. The observation leading to this opinion has to do with PC- fractions, as in Example 3. There the reference continued fraction had all $\beta_n = 0$. If instead of that we only require $\beta_{2k+1} = 0$, $\beta_{2k} = \beta$, $k = 1, 2, 3, \ldots$, we get the continued fraction

$$\frac{1}{2} - \frac{1}{1} + \frac{1}{\beta z} + \frac{z}{0} + \frac{1}{\beta z} + \frac{z}{0} + \cdots.$$

It is easily seen, that this is the PC-expansion of the pair

$$\left(\frac{1}{2} - \frac{\beta z}{1 - (1 - \beta)z}, -\frac{1}{2} \right).$$

Will nearness as in Example 3 of $(L_o(z), L_\infty(z))$ to this particular pair imply existence of a representing (corresponding) PC-fraction, and will it in the sense of Example 3 be near the one with $\beta_{2k+1} = 0$, $\beta_{2k} = \beta$? We shall see that at least the last part of the question sometimes has a negative answer. We shall in the following restrict the considerations to the case

$$\beta = 1,$$

which has the advantage that there will be no pole to worry about in the first reference function. With another choice of β we run into difficulties in trying to find the proper type of nearness to be inherited in the recursion process. Moreover, the nearness condition will fit exactly into the conditions in Example 1, except for normalizations. The reference pair is in this case

$$\left(\frac{1}{2} - z, -\frac{1}{2} \right),$$

and it is represented by the PC-fraction

$$\frac{1}{2} - \frac{1}{1} + \frac{1}{z} + \frac{z}{0} + \frac{1}{z} + \frac{z}{0} + \cdots .$$

A slight change of normalization in the result of Example 1 shows that if

$$L_o(z) = \frac{1}{2} - z + c_2 z^2 + \cdots ,$$

$$L_\infty(z) = -\frac{1}{2} - c_{-1} z^{-1} - c_2 z^{-2} - \cdots ,$$

are such that L_o is holomorphic and

$$\left| L_o(z) - \left(\frac{1}{2} - z \right) \right|$$

is sufficiently small in a large enough disk $|z| < R$, and such that L_∞ is holomorphic and

$$\left| L_\infty(z) + \frac{1}{2} \right|$$

is sufficiently small in a large enough domain $|z| > \rho$, then there exists a corresponding general T-fraction

$$\frac{1}{2} - \frac{z}{1+z} + \frac{F_2 z}{1+G_2 z} + \frac{F_3 z}{1+G_3 z} + \cdots ,$$

where, for some $F \neq 0$

$$\lim_{n \to \infty} F_n = F$$

exists, and

$$\lim_{n \to \infty} G_n = -F.$$

Remark. The phrazes "sufficiently small" and "large enough" may here be replaced by precisely stated conditions from [5]. Such reformulations will not be needed for what we here want to draw attention to.

From [2, Thm 4.1] we know that our T-fraction is the even part of a PC-fraction, where

$$\beta_{2k} = G_1 G_2 \cdots G_k, k \geq 1.$$

$$\beta_{2k+1} = \left(1 + \frac{F_{k+1}}{G_{k+1}} \right) \frac{1}{G_1 G_2 \cdots G_k}, k \geq 1.$$

(For correspondence of general T-fractions at ∞, $G_n \neq 0$ is necessary [6].) Hence, if a corresponding PC-fraction exists, it must have this particular form, where we regard all F_n and G_n to be known. It turns out, that the remaining correspondence property (i. e. the odd part correspondence) is easily established, and we have the observation:

Observation 1. *Let*

$$L_o(z) = \frac{1}{2} - z + c_2 z^2 + c_3 z^3 + \cdots,$$

$$L_\infty(z) = -\frac{1}{2} - c_{-1} z^{-1} - c_{-2} z^{-2} - \cdots$$

be holomorphic in large enough neighborhoods $|z| < R$ and $|z| > \rho$ of 0 and ∞ respectively, and

$$\left| L_o(z) - \left(\frac{1}{2} - z \right) \right|, \left| L_\infty(z) + \frac{1}{2} \right|$$

sufficiently small in those neighborhoods. Then there exists a corresponding PC-fraction

$$\frac{1}{2} - \frac{1}{1} + \frac{1}{\beta_2 z} + \frac{(1 - \beta_2 \beta_3)z}{\beta_3} + \frac{1}{\beta_4 z} + \frac{(1 - \beta_4 \beta_5)z}{\beta_5} + \cdots,$$

where $\beta_2 = 1, \beta_4 = G_2, \beta_6 = G_2 G_3, \ldots, \beta_{2k} = G_2 G_3 \cdots G_k$, and

$$\beta_3 = 1 + \frac{F_2}{G_2}, \beta_5 = \left(1 + \frac{F_3}{G_3} \right) \frac{1}{G_2}, \cdots,$$

$$\beta_{2k+1} = \left(1 + \frac{F_{k+1}}{G_{k+1}} \right) \frac{1}{G_2 G_3 \cdots G_k},$$

and where $F_n \to F \neq 0, G_n \to -F$ when $n \to \infty$.

Remark. This observation illustrates the importance of the reference pair. The same type of nearness to $(1/2, -1/2)$ and to $(1/2 - z, -1/2)$, i. e. small differences in large domains, leads to very different conclusions as far as corresponding PC-fraction is concerned, i. e. limit periodicity in one case and (generally) not in the other one. But in both cases we can conclude existence. We shall look a little more closely at the possible cases:

We have

$$1 - \beta_{2k} \beta_{2k+1} = -\frac{F_{k+1}}{G_{k+1}} \to 1$$

when $k \to \infty$. It is known, that $F_n \to F$ and $G_n \to -F$ geometrically under the conditions in Observation 1. Hence, if

$$F_n \to -1,$$

we have $\beta_{2k} \to$ limit $(\neq 0)$, in which case $\beta_{2k+1} \to 0$. In this *very special* case we have limit periodicity. In the other cases we have:

$|F| = 1, F \neq -1.$

$$\lim_{k \to \infty} \beta_{2k}$$

does not exist,

$$\lim_{k \to \infty} |\beta_{2k}|$$

exists and is $\neq 0$. Finally

$$\lim_{k \to \infty} \beta_{2k+1} = 0.$$

$|F| < 1.$

$$\lim_{k \to \infty} \beta_{2k} = 0.$$

Nothing can be said about β_{2k+1} *generally* in this case.

$|F| > 1$.

$$\lim_{k\to\infty} \beta_{2k} = \infty, \ \lim_{k\to\infty} \beta_{2k+1} = 0.$$

These cases are all illustrated in the following trivial example:

Example 4. If

$$(L_o(z), L_\infty(z)) = \left(\frac{1}{2} - \frac{z}{1 + (a+1)z}, -\frac{1}{2}\right),$$

the nearness condition in Observation 1 is satisfied for all a sufficiently close to -1. (Again we refer to [5, Sec. 2] for a precise definition of nearness in this case.) The corresponding general T-fraction is

$$\frac{1}{2} + \frac{-z}{1+z} + \frac{az}{1-az} + \frac{az}{1-az} + \cdots,$$

and the corresponding PC-fraction is given by

$$\beta_{2k} = (-a)^{k-1}, \beta_{2k+1} = 0, k \geq 1.$$

Here $F = a$. In any neighborhood of $a = -1$ there are values of $|a|$ that are $< 1, = 1$ and > 1.

So far our examples have only served to illustrate the significance of the reference function(s). The next example will illustrate the role of the nearness definition.

Example 5. Take an arbitrary formal power series of the form

$$L_o(z) = \frac{1}{2} + c_1 z + c_2 z^2 + c_3 z^3 + \cdots$$

and let $L_\infty(z)$ be the particular series $-\frac{1}{2}$. Then it is readily seen, that a unique corresponding PC-fraction exists: With

$$-\frac{c_1 z + c_2 z^2 + c_3 z^3 + \cdots}{1 + c_1 z + c_2 z^2 + c_3 z^3 + \cdots} = \beta_2 z + \beta_4 z^2 + \beta_6 z^3 + \cdots . \diamond B$$

we have

$$c_1 z + c_2 z^2 + c_3 z^3 + \cdots = \frac{-\beta_2 z - \beta_4 z^2 - \beta_6 z^3 - \cdots}{1 + \beta_2 z + \beta_4 z^2 + \beta_6 z^3 + \cdots} =$$

$$= \frac{-1}{1 + \dfrac{1}{\beta_2 z + \beta_4 z^2 + \cdots}},$$

and the pair

$$\left(\frac{1}{2} + c_1 z + c_2 z^2 + c_3 z^3 + \cdots, -\frac{1}{2}\right)$$

may be written

$$\left(\frac{1}{2} - \frac{1}{1 + \dfrac{1}{\beta_2 z + \beta_4 z^2 + \cdots}}, -\frac{1}{2}\right).$$

It is easily checked, that this pair is represented by the PC-fraction

$$\frac{1}{2} - \frac{1}{1} + \frac{1}{\beta_2 z} + \frac{z}{0} + \frac{1}{\beta_4 z} + \frac{z}{0} + \cdots .$$

(this holds for any sequence $\{\beta_{2k}\}$ of complex numbers, also e. g. for the sequence $0, 0, 0, \ldots .)$

A function "near" $L_o(z)$ is in this case obtained by a perurbation of the "β-series", as described in Observation 2. This leads to the following trivial observation:

Observation 2. Let $B(z)$ and $\phi(z)$ be the formal series

$$B(z) = \beta_2 z + \beta_4 z^2 + \beta_6 z^3 + \cdots$$

and

$$\phi(z) = \varepsilon_1 z + \varepsilon_2 z^2 + \varepsilon_3 z^3 + \cdots ,$$

where $\varepsilon_n \to 0$ when $n \to \infty$. Then

$$\left(\frac{1}{2} - \cfrac{1}{1 + \cfrac{1}{B(z) + \phi(z)}} , -\frac{1}{2} \right)$$

has a corresponding PC-fraction

$$\frac{1}{2} - \frac{1}{1} + \frac{1}{\beta_2^* z} + \frac{z}{0} + \frac{1}{\beta_4^* z} + \frac{z}{0} + \frac{1}{\beta_6^* z} + \cdots ,$$

where $\beta_n^* - \beta_n \to 0$ when $n \to \infty$.

Remark. Most interesting are perhaps the cases where the formal power series represent analytic functions. The simplest case when $B(z)$ represents an analytic function is when all β_{2k} are equal to a fixed β, in which case

$$B(z) = \frac{\beta z}{1 - z}.$$

A function "near" B is here in particular obtained by adding any function ϕ, holomorphic in $|z| < R$ for some $R > 1$, and with $\phi(0) = 0$.

Final remark. A similar discussion may be carried out by requiring, in the reference PC-fraction, all even order β to be 0, and all odd order β to have the fixed value 1. In this case the reference pair is $(1/2, -1/2 + 1/z)$. By using M-fraction correspondence we are led to an observation closely related to Observation 1. Also an observation related to Observation 2 can be made.

References

1. L. Jacobsen and H. Waadeland, *A result on nearness of functions and their regular C-fraction expansions*, Proc. of the Amer. Math. Soc. 105 **4** (1989).
2. W. B. Jones, O. Njåstad and W. J. Thron, *Continued fractions associated with the trigonometric and other strong moment problems*, Constructive Approximation **2** (1986), 197-211.
3. W. B. Jones and W. J. Thron, "Continued Fractions: Analytic Theory and Applications. Encyclopedia Math. Appl. 11.," Addison-Wesley Publ. Co., Reading, Mass., distributed now by Cambridge University Press, New York, 1980.
4. O. Njåstad and H. Waadeland, *Strongly bounded functions and corresponding PC-fractions*, (to appear).
5. H. Waadeland, *General T-fractions corresponding to functions satisfying certain boundedness conditions*, Journal of Approximation Theory **26** (1979), 317-328.
6. H. Waadeland, *Some properties of general T-fractions*, Dept. of Mathematics, University of Trondheim, Norway **5** (1978).

Continued fraction identities derived from the
invariance of the crossratio under *l.f.t.*

W. J. Thron*

1. Introduction. The *crossratio* of four distinct complex numbers u, v, w, z is defined to be the fraction

$$\left(\frac{u-v}{u-w}\right)\left(\frac{w-z}{v-z}\right).$$

It shall be denoted by

(1.1)
$$[u, v, w, z] := \left(\frac{u-v}{u-w}\right)\left(\frac{w-z}{v-z}\right).$$

If either of the numbers, say z, equals ∞ we have

$$[u, v, w, \infty] = \frac{u-v}{u-w}$$

The crossratio is invariant under a non-singular linear fractional transformation $(\ell.f.t.)T$.

(1.2)
$$[T(u), T(v), T(w), T(z)] = [u, v, w, z].$$

This invariance property characterizes $\ell.f.t.$ in the sense that if a mapping $T : \mathbf{C} \to \mathbf{C}$ satisfies (1.2) for $u \in \mathbf{C}$ and fixed v_0, w_0, z_0, then it is an $\ell.f.t.$

If the order of the elements of the crossratio is changed, the crossratio may remain unchanged. This is the case for

(1.3)
$$[u, v, w, z] = [z, w, v, u] = [v, u, z, w] = [w, z, u, v].$$

Under all other permutations the crossratio changes its value. Thus, if $\lambda = [u, v, w, z]$, then

(1.4)
$$[u, w, v, z] = 1/\lambda =: \sigma(\lambda),$$

and

(1.5)
$$[u, z, w, v] = 1 - \lambda =: \tau(\lambda).$$

The other possible values are obtained from composition of σ and τ and are the following:

$$\sigma \circ \tau(\lambda) = \frac{1}{1-\lambda}, \quad \tau \circ \sigma(\lambda) = 1 - \frac{1}{\lambda} = \frac{\lambda-1}{\lambda}$$

$$\sigma \circ \tau \circ \sigma(\lambda) = \frac{\lambda}{\lambda-1}.$$

*This research was supported in part by the U.S. National Science Foundation under Grant No. DMS-8700498.

Note that σ and τ are both idempotent, that is $\sigma \circ \sigma(\lambda) = \lambda$, $\tau \circ \tau(\lambda) = \lambda$.

Thus there are altogether six values (which may all be distinct but need not be) which a cross-ratio of four given elements can take on under permutation of these elements.

The formula (1.2) can be applied to continued fractions since they can be defined in terms of $\ell.f.t.$ as follows:

$$s_n(w) := \frac{a_n}{b_n + w}, \qquad a_n \neq 0,$$

$$S_n(w) := S_{n-1}(s_n(w)), \quad n \geq 2, \quad S_1(w) := s_1(w).$$

The nth $approximant$ of the continued fraction $K(a_n/b_n)$ is the quantity $f_n := S_n(0)$. For modified continued fractions $K(a_n, b_n, w_n)$ we have $\{S_n(w_n)\}$ as the $sequence\ of\ approximants$. We further introduce

$$S_k^{(n)}(w) := s_{n+1} \circ \cdots \circ s_{n+k}(w), \quad f_k^{(n)} := S_k^{(n)}(0),$$

$$f := \lim_{n \to \infty} S_n(0), \quad f^{(n)} := \lim_{k \to \infty} S_k^{(n)}(0),$$

provided the limits exist. The quantity $f^{(n)}$ is called the nth $tail$ of $K(a_n/b_n)$. The following recursion relation holds

$$(1.6) \qquad\qquad f^{(n)} = -b_n + \frac{a_n}{f^{(n-1)}} = s_n^{-1}(f^{(n-1)}).$$

It is also clear that

$$f = S_n(f^{(n)}) \quad \text{or} \quad f^{(n)} = S_n^{-1}(f).$$

Any sequence $\{g^{(n)}\}$ satisfying (1.6) or equivalently

$$g^{(n)} = S_n^{-1}(g), \quad n \leq 1,$$

will be called a $tail\ sequence$ of $K(a_n/b_n)$. After $\{f^{(n)}\}$ the most important tail sequences are $\{S_n^{-1}(\infty)\}$ and $\{S_n^{-1}(0)\}$. The first is known as the $critical\ tail\ sequence$. We set

$$h_n := -S_n^{-1}(\infty), \quad k_n := -S_n^{-1}(0)$$

and thus have

$$(1.6') \qquad\qquad h_n = b_n + \frac{a_n}{h_{n-1}}, \quad h_1 = b_1$$

and

$$(1.6'') \qquad\qquad k_n = b_n + \frac{a_n}{k_{n-1}}, \quad k_1 = \infty.$$

The following formulas will be very useful in the sequel

$$(1.7) \qquad
\begin{aligned}
&\text{(a)} \quad S_n(\infty) = S_{n-1}(s_n(\infty)) = S_{n-1}(0) = f_{n-1}, \\
&\text{(b)} \quad S_n(-b_n) = S_{n-1}(s_n(-b_n)) = S_{n-1}(\infty) = f_{n-2}, \\
&\text{(c)} \quad S_n(a_{n+1}/b_{n+1}) = S_n(s_{n+1}(0)) = S_{n+1}(0) = f_{n+1}.
\end{aligned}$$

To obtain formulas, of importance in the theory of continued fractions, from (1.2) the choice

$$T = S_n$$

appears to be best. Even formulas for tails $S_n^{-1}(g)$ can be obtained in this way, though in that case, the choice $T = S_n^{-1}$ may be simpler and more suggestive. We begin with

(1.2') $$[S_n(u_n), S_n(v_n), S_n(w_n), S_n(z_n)] = [u_n, v_n, w_n, z_n].$$

Different formulas will be obtained if a different set $\{u_n, v_n, w_n, z_n\}$ of elements is chosen. But under a permutation of the elements of the same set an identity $A = B$ is replaced by

$$\chi(A) = \chi(B),$$

where χ is a composition of transformations σ and τ. An identity $\chi(A) = \chi(B)$ is *equivalent* to the initial $A = B$, but this equivalence is frequently hard to recognize in particular if $\chi = \tau$ or a transformation involving τ. Thus we shall list some equivalent formulas separately since they look quite different.

Many of the formulas we are about to derive were originally only obtained for the case $K(a_n/1)$. We shall give credit to the author who, to the best of our knowledge, first proved one of these identities even if it was done only for continued fractions $K(a_n/1)$. The older formulas were not derived from (1.2') but computed directly.

The purpose of the present article is to present a unified approach to the known identities as well as to give a systematic exploration leading to other useful identities.

In [2] formulas that can be derived from (1.2') (but were not) play an important role. In [3] the "invariance of the cross-ratio" is explicitly appealed to (possibly for the first time) as the underlying principle for the formulas used. In the survey article [11] the invariance of the cross-ratio was again emphasized as the source of useful identities, which were then applied throughout the paper.

2. The earliest application. In 1934 Wall [13] (see also [5, p. 127]) derived the formulas

(2.1) $$[f_{n-2}, f_{n-1}, f_n, f_{n+1}] = -\frac{a_{n+1}}{b_n b_{n+1}}.$$

It can be obtained, using (1.7), by setting $u_n = -b_n, v_n = \infty, w_n = 0, z_n = a_{n+1}/b_{n+1}$ in (1.2'). As an illustration of other formulas involving four non-successive approximants we mention

(2.2) $$[f_{n-1}, f_n, f_{n+1}, f_{n+3}] = -\frac{b_n \left(\frac{a_{n+1}}{b_{n+1}} - f_3^{(n)}\right)}{\left(b_n + \frac{a_{n+1}}{b_{n+1}}\right) f_3^{(n)}},$$

which is obtained by setting $u_n = -b_n, v_n = 0, w_n = a_{n+1}/b_{n+1}, z_n = f_3^{(n)}$.

3. The substitution $z_n = -h_n$. The identities that can be derived from (1.2') are substantially simplified if one chooses $z_n = -h_n$ since this leads to $S_n(z_n) = \infty$ and hence

(3.1) $$\frac{S_n(u_n) - S_n(v_n)}{S_n(u_n) - S_n(w_n)} = \left(\frac{u_n - v_n}{u_n - w_n}\right) \left(\frac{w_n + h_n}{v_n + h_n}\right).$$

This formula occurs in [9, p. 129]. For its validity it is, as was pointed out earlier, necessary that u_n, v_n, w_v and $-h_n$ be all distinct numbers in the extended complex plane. Similar restrictions hold for all formulas in the remainder of this article. They will not be repeated in each instance.

The choice $u_n = f^{(n)}$, $w_n = 0$ then results in

$$(3.2) \qquad \frac{f - S_n(v_n)}{f - f_n} = \frac{h_n}{f^{(n)}} \left(\frac{f^{(n)} - v_n}{h_n + v_n} \right),$$

which was first derived by Thron and Waadeland [10] for the study of acceleration of convergence.

If certain conditions on the elements a_n, b_n of the continued fraction $K(a_n, b_n)$ are known to hold only from some $N + 1$ on the identity

$$(3.3) \qquad \frac{f - S_{N+k}(w_{N+k})}{f - f_{N+k}} = \left(\frac{f^{(N)} - S_k^{(N)}(w_{N+k})}{f^{(N)} - f_k^{(N)}} \right) \left(\frac{f_k^{(N)} + h_N}{S_k^{(N)}(w_{N+k}) + h_N} \right)$$

may be useful. Note that all elements occurring in the first parenthesis on the right satisfy the conditions. Only h_N involves a_n, b_n with $n \leq N$.

From (3.1) one obtains by choosing $u_n = \infty$ or $w_n = \infty$, two essentially different looking simplifications of (3.1). They are

$$(3.4a) \qquad \frac{f_{n-1} - S_n(v_n)}{f_{n-1} - S_n(w_n)} = \frac{w_n + h_n}{v_n + h_n}$$

and

$$(3.4b) \qquad \frac{S_n(u_n) - S_n(v_n)}{S_n(u_n) - f_{n-1}} = \frac{v_n - u_n}{v_n + h_n}.$$

However the two formulas are equivalent since τ transforms (3.4a) into (3.4b).

Substituting $v_n = f^{(n)}$, $w_n = 0$ into (3.4a) one arrives at

$$(3.5) \qquad \frac{f_{n-1} - f}{f_{n-1} - f_n} = \frac{h_n}{f^{(n)} + h_n},$$

which can be found in [11, formula (3.1)] while $v_n = 0$, $w_n = -b_n$ gives

$$\frac{f_{n-1} - f_n}{f_{n-1} - f_{n-2}} = \frac{-b_n + h_n}{h_n}.$$

Writing it as

$$(3.6) \qquad f_n - f_{n-1} = \left(\frac{b_n - h_n}{h_n} \right) (f_{n-1} - f_{n-2})$$

is more suggestive. In this form it is due to Overholt [8].

The substitution $v_n = 0$, $w_n = a_{n+1}/b_{n+1}$ results in

$$\frac{f_{n-1} - f_n}{f_{n-1} - f_{n+1}} = \frac{a_{n+1} + b_{n+1} h_n}{b_{n+1} h_n}$$

which in the form

(3.7)
$$f_n - f_{n-1} = \left(\frac{b_{n+1} h_n + a_{n+1}}{b_{n+1} h_n} \right) (f_{n+1} - f_{n-1})$$

was used by Lange [7] to deduce convergence of the whole continued fraction if its even or odd part is known to converge.

Replacing u_n by $f^{(n)}$ in (3.4b) one arrives at

(3.8)
$$\frac{f - S_n(v_n)}{f - f_{n-1}} = \frac{v_n - f^{(n)}}{v_n + h_n},$$

which could be used instead of (3.2) in the analysis of convergence acceleration and is somewhat simpler. As we shall see in other cases using $f_{n-1} = S_n(\infty)$ instead of $f_n = S_n(0)$ usually leads to somewhat simpler formulas. If it is not known whether the continued fraction converges the substitution $u_n = f_k^{(n)}$, which leads to

(3.9)
$$\frac{f_{n+k} - S_n(v_n)}{f_{n+k} - f_{n-1}} = \frac{v_n - f_k^{(n)}}{v_n + h_n},$$

may be useful.

If in (3.8) one sets $v_n = 0$, one obtains

(3.10)
$$\frac{f - f_n}{f - f_{n-1}} = -\frac{f^{(n)}}{h_n},$$

which occurs in Jacobsen and Waadeland [4, p. 101, formula 15].

The formula of Hayden [1]

(3.11)
$$\frac{f_n - f}{f_n - f_{n-1}} = \frac{f^{(n)}}{f^{(n)} + h_n}$$

is obtained from (3.4b) by setting $u_n = 0$, $v_n = f^{(n)}$. The formula can also be derived by applying the transformation $\sigma \circ \tau$ to (3.10). If the convergence of the continued fraction has not as yet been established, one may prefer the identity

(3.12)
$$\frac{f_n - f_{n+k}}{f_n - f_{n-1}} = \frac{f_k^{(n)}}{f_k^{(n)} + h_n}$$

in which $u_n = 0$, $v_n = f_k^{(n)}$.

If one sets $u_n = 0$, $v_n = a_{n+1}/b_{n+1}$ in (3.4b) one gets

(3.13)
$$\frac{f_n - f_{n+1}}{f_n - f_{n-1}} = \frac{a_{n+1}}{a_{n+1} + h_n b_{n+1}}.$$

It is related to the formula (3.6) by replacing n by $n - 1$ and using (1.6'). It can be found in [1].

The choice $u_n = a_{n+1}/b_{n+1}$, $v_n = 0$ leads to a second formula of Lange [7]

(3.14)
$$f_{n+1} - f_n = \frac{-a_{n+1}}{b_{n+1} h_n} (f_{n+1} - f_{n-1})$$

The two formulas are equivalent (the transformation is τ).

4. Identities for tails.

Replacing T by S_n^{-1} in (1.2) one obtains

(4.1)
$$\left[S_n^{-1}(u_n), S_n^{-1}(v_n), S_n^{-1}(w_n), S_n^{-1}(z_n)\right] = [u_n, v_n, w_n, z_n].$$

Clearly this formula is equivalent to (1.2') as one can see by setting $u_n = S_n(u'_n)$, $v_n = S_n(v'_n)$, in (4.1). However (4.1) provides a better starting point for this section since the identities that we are seeking here are concerned with comparisons of different tail sequences and thus differ markedly from the identities studied in Section 3.

Setting $z_n = \infty$ in (4.1) we arrive at

(4.2)
$$\left(\frac{S_n^{-1}(u_n) - S_n^{-1}(v_n)}{S_n^{-1}(u_n) - S_n^{-1}(w_n)}\right)\left(\frac{S_n^{-1}(w_n) + h_n}{S_n^{-1}(v_n) + h_n}\right) = \frac{u_n - v_n}{u_n - w_n}.$$

The choice $u_n = f$, $v_n = g$, $w_n = p$ leads to

(4.3)
$$\left(\frac{f^{(n)} - g^{(n)}}{f^{(n)} - p^{(n)}}\right)\left(\frac{p^{(n)} + h_n}{g^{(n)} + h_n}\right) = \frac{f - g}{f - p},$$

or with $u_n = g$, $v_n = f$, $w_n = p$ to the equivalent identity

(4.4)
$$\left(\frac{g^{(n)} - f^{(n)}}{g^{(n)} - p^{(n)}}\right)\left(\frac{p^{(n)} + h_n}{f^{(n)} + h_n}\right) = \frac{g - f}{g - p}.$$

If in addition to $z_n = \infty$ we choose $w_n = f_{n-1}$ in (4.1), we obtain

(4.5)
$$\frac{S_n^{-1}(u_n) - S_n^{-1}(v_n)}{S_n^{-1}(v_n) + h_n} = \frac{v_n - u_n}{u_n - f_{n-1}}.$$

For $u_n = g$, $v_n = p$ one gets

(4.6)
$$\frac{g^{(n)} - p^{(n)}}{p^{(n)} + h_n} = \frac{p - g}{g - f_{n-1}}.$$

The choices $u_n = f$, $v_n = g$ or $u_n = g$, $v_n = f$ lead to

(4.7)
$$\frac{f^{(n)} - g^{(n)}}{g^{(n)} + h_n} = \frac{g - f}{f - f_{n-1}}$$

and

(4.8)
$$\frac{g^{(n)} - f^{(n)}}{f^{(n)} + h_n} = \frac{f - g}{g - f_{n-1}},$$

respectively.

Equivalent formulas, but quite different in appearance, result from the substitution $u_n = f_{n-1}$, $z_n = \infty$ in (4.1). We have

(4.9)
$$\frac{S_n^{-1}(w_n) + h_n}{S_n^{-1}(v_n) + h_n} = \frac{f_{n-1} - v_n}{f_{n-1} - w_n},$$

and hence

$$(4.10) \qquad \frac{g^{(n)} + h_n}{p^{(n)} + h_n} = \frac{f_{n-1} - p}{f_{n-1} - g},$$

and

$$(4.11) \qquad \frac{f^{(n)} + h_n}{g^{(n)} + h_n} = \frac{f_{n-1} - g}{f_{n-1} - f}.$$

The choice $w_n = f_n$, $z_n = \infty$ leads to considerably more complicated formulas than the ones listed above and so we shall not give them.

For $z_n = f_{n-1}$ (4.1) becomes

$$(4.12) \qquad \frac{S_n^{-1}(u_n) - S_n^{-1}(v_n)}{S_n^{-1}(u_n) - S_n^{-1}(w_n)} = \left(\frac{u_n - v_n}{u_n - w_n} \right) \left(\frac{w_n - f_{n-1}}{v_n - f_{n-1}} \right).$$

The substitution $u_n = g$, $v_n = k$, $w_n = m$ then leads to

$$(4.13) \qquad \frac{g^{(n)} - p^{(n)}}{g^{(n)} - m^{(n)}} = \left(\frac{g - p}{g - m} \right) \left(\frac{m - f_{n-1}}{k - f_{n-1}} \right).$$

Other substitutions yield

$$(4.14) \qquad \frac{f^{(n)} - g^{(n)}}{f^{(n)} - p^{(n)}} = \left(\frac{f - g}{f - p} \right) \left(\frac{f - f_{n-1}}{g - f_{n-1}} \right)$$

and

$$(4.15) \qquad \frac{g^{(n)} - f^{(n)}}{g^{(n)} - p^{(n)}} = \left(\frac{g - f}{g - p} \right) \left(\frac{p - f_{n-1}}{f - f_{n-1}} \right).$$

Clearly, some of the earlier formulas are obtained by setting g, k or $m = \infty$.

The choice $u_n = f_n$, $z_n = f_{n-1}$ leads to

$$(4.16) \qquad \frac{S_n^{-1}(v_n)}{S_n^{-1}(w_n)} = \left(\frac{f_n - v_n}{f_n - w_n} \right) \left(\frac{w_n - f_{n-1}}{v_n - f_{n-1}} \right).$$

With $v_n = f$, $w_n = g$ one arrives at

$$(4.17) \qquad \frac{f^{(n)}}{g^{(n)}} = \left(\frac{f_n - f}{f_n - g} \right) \left(\frac{g - f_{n-1}}{f - f_{n-1}} \right).$$

Setting $g = \infty$ one obtains (3.10).

Results of the substitution u_n, v_n, w_n or $z_n = 0$ which amounts to introducing $k_n = -S_n^{-1}(0)$, will be considered in the next section.

5. The substitution $u_n = S_n^{-1}(0)$. Of some interest, though certainly less important than the substitution $z_n = -h_n$ is the substitution $u_n = -k_n$ in (1.2′). It yields

$$(5.1) \qquad \frac{S_n(v_n)}{S_n(w_n)} \left(\frac{S_n(w_n) - S_n(z_n)}{S_n(v_n) - S_n(z_n)} \right) = \left(\frac{k_n + v_n}{k_n + w_n} \right) \left(\frac{w_n - z_n}{v_n - z_n} \right).$$

We know of no worthwhile application of this formula other than making the additional substitution $z_n = -h_n$ which leads to

(5.2)
$$\frac{S_n(v_n)}{S_n(w_n)} = \left(\frac{k_n + v_n}{k_n + w_n}\right)\left(\frac{w_n + h_n}{v_n + h_n}\right).$$

This is further simplified by setting $v_n = \infty$ to

(5.3)
$$\frac{f_{n-1}}{S_n(w_n)} = \frac{w_n + h_n}{w_n + k_n}.$$

To obtain additional formulas for tail sequences one can set $u_n = 0$ in (4.2) to get

(5.4)
$$\left(\frac{k_n + S_n^{-1}(v_n)}{k_n + S_n^{-1}(w_n)}\right)\left(\frac{S_n^{-1}(w_n) + h_n}{S_n^{-1}(v_n) + h_n}\right) = \frac{v_n}{w_n},$$

which is of course equivalent to (5.2). The substitutions $v_n = g$, $w_n = m$ then yield

(5.5)
$$\left(\frac{k_n + g^{(n)}}{k_n + m^{(n)}}\right)\left(\frac{m^{(n)} + h_n}{g^{(n)} + h_n}\right) = \frac{g}{m}.$$

6. Composite formulas. In [5, p. 307, formula (8.3.18)] (also in [12]) we find the identity

(6.1)
$$\frac{f - S_n(u_n)}{f_{n-1} - f_n} = \frac{h_n(f^{(n)} - u_n)}{(h_n + u_n)(h_n + f^{(n)})},$$

which differs essentially from any of the formulas we have encountered heretofore in that it has four distinct expressions on the left. To obtain it one can proceed as follows. In (3.1) we change the notation by setting $u_n = w_n$, $v_n = u_n$, $w_n = f_n$ to arrive at

(6.2)
$$\frac{S_n(w_n) - S_n(u_n)}{S_n(w_n) - S_n(t_n)} = \left(\frac{w_n - u_n}{w_n - t_n}\right)\left(\frac{t_n + h_n}{u_n + h_n}\right).$$

multiplication of (3.1) by (6.2) then yields

(6.3)
$$\frac{S_n(u_n) - S_n(v_n)}{S_n(w_n) - S_n(t_n)} = \left(\frac{u_n - v_n}{w_n - t_n}\right)\left(\frac{w_n + h_n}{u_n + h_n}\right)\left(\frac{t_n + h_n}{v_n + h_n}\right)$$

from which (6.1) can be derived by choosing $v_n = f^{(n)}$, $w_n = \infty$, $t_n = 0$.

Another composite formula can be obtained from (3.5). We first write it as

(6.4)
$$f - f_{n-1} = \frac{h_n}{f^{(n)} + h_n}(f_n - f_{n-1}).$$

Iterating (3.6) leads to

$$f_n - f_{n-1} = \left(\frac{a_1}{b_1}\right)\prod_{m=2}^{n}\left(\frac{b_m - h_m}{h_m}\right).$$

Combining this with (6.4) one arrives at

(6.5)
$$f - f_{n-1} = \frac{a_1 h_n}{b_1(f^{(n)} + h_n)}\prod_{m=2}^{n}\left(\frac{b_m - h_m}{h_m}\right),$$

which can be found in [11].

7. Formulas in terms of chordal distance.

The chordal distance between two points $u, v \in \hat{C}$ is defined as

$$d(u, v) = \frac{|u - v|}{\sqrt{1 - |u|^2}\sqrt{1 + |t|^2}}, \quad d(u, \infty) = \frac{1}{\sqrt{1 + |u|^2}}$$

Thus if u, v, w, z are all finite complex numbers, then

$$|[u, v, w, z]| = \frac{d(u, v)d(w, z)}{d(u, w)d(v, z)}.$$

From (1.2') one thus obtains

(7.1)
$$\frac{d(S_n(u_n), S_n(v_n))d(S_n(w_n), S_n(z_n))}{d(S_n(u_n), S_n(w_n))d(S_n(v_n), S_n(z_n))} = \frac{d(u_n, v_n)d(w_n, z_n)}{d(u_n, w_n)d(v_n, z_n)},$$

which is formula (1.7) in [3] and the formula following (3.13) in [2]. This formula is also used in the proofs of Theorems 3.6 and 3.9 in [6].

For the analogue of (3.1) we get the formula

(7.2)
$$\frac{d(S_n(u_n), S_n(v_n))}{d(S_n(u_n), S_n(w_n))}\sqrt{\frac{1 + |S_n(v_n)|^2}{1 + \sqrt{|S_n(w_n)|^2}}} = \frac{d(u_n, v_n)d(w_n, -h_n)}{d(u_n, w_n)d(v_n, -h_n)}.$$

8. Addenda.

The referee kindly pointed out a number of additional formulas that can be derived from the invariance of the cross ratio and have appeared in the literature.

Hayden [1] besides (3.11) and (3.13) has the following formulas:

(8.1)
$$\frac{S_n(w) - S_n(z)}{f_{n-1} - f_n} = \frac{h_n(w - z)}{(w + h_n)(z + h_n)}$$

and

(8.2)
$$\frac{S_n(w) - f_n}{f_n - f_{n-1}} = \frac{-w}{h_n + w}.$$

The formula (8.1) can be obtained from (6.3) by setting $u_n = w$, $v_n = z$, $w_n = \infty$, $A_n = 0$. Letting $k = 0$, $v_n = w$ in (3.9) one arrives at (8.2).

Phipps [16] has the formula

(8.3)
$$\frac{S_n(w_n) - f_{n-1}}{f_n - f_{n-1}} = \frac{h_n}{h_n + w_n},$$

which follows from (3.4a)

Niethammer and Wietschorke [15] have the identity

$$\frac{S_n(w) - S_{n-1}(w)}{S_{n-1}(w) - S_{n-2}(w)} = \frac{-a_{n-1}(h_{n-2} + w)(a_n - w - w^2)}{h_{n-1}h_{n-2}(h_n + w)(a_{n-1} - w - w^2)}.$$

Using (3.1) with $u_n = s_n^{-1}(w) = (a_n - w)/w$, $v_n = w$, $w_n = s_n^{-1} \circ s_{n-1}^{-1}(w) = (a_n w = a_{n-1} + w)/(a_{n-1} - w)$

we arrive at

$$(8.4) \qquad \frac{S_n(w) - S_{n-1}(w)}{S_{n-1}(w) - S_{n-2}(w)} = \frac{-(a_n - w - w^2)(a_n w - a_{n-1} + w + h_n(a_{n-1} - w))}{a_n(a_{n-1} - w - w^2)(w + h_n)},$$

which is somewhat simpler but presumably equivalent to their expression.

To the formulas in Section 4 one can add

$$(8.5) \qquad \frac{S_n(w_n) - f}{f - g} = \left(\frac{w_n - f^{(n)}}{f^{(n)} - g^{(n)}} \right) \cdot \left(\frac{h_n + g^{(n)}}{h_n + w_n} \right),$$

which is given in [2].

Another application of (3.1) is to obtain the canonical form for S_n, provided it has two distinct fixed points v_n and w_n. Writing $u_n = u$ this leads to

$$(8.6) \qquad \frac{S_n(u) - v_n}{S_n(u) - w_n} = \left(\frac{u - v_n}{u - w_n} \right) \left(\frac{w_n + h_n}{v_n + h_n} \right), \quad v_n \neq w_n \text{ are the fixed points of } S_n.$$

Finally, another example of iteration that is given by Jacobsen in [14], in a more general setting, can be obtained as follows. In (1.2) let $T = s_n^{-1}$, $u = S_{n-1}^{-1}(\infty) = -h_{n-1}$, $v = S_{n-1}^{-1}(p) = p^{n-1)}$, $w = S_{n-1}^{-1}(g) = g^{(n-10}$, $z = s_n(\infty) = 0$. Then

$$(8.7) \qquad \begin{aligned} \frac{h_n - g^{(n)}}{h_n - p^{(n)}} &= \frac{s_n^{-1}(-h_{n-1}) + s_n^{-1}(g^{(n-1)}}{s_n^{-1}(-h_{n-1}) + s_n^{-1}(p^{(n-1)}} \\ &= \left(\frac{h_{n-1} - g^{(n-1)}}{h_{n-1} - p^{(n-1)}} \right) \cdot \left(\frac{p^{(n-1)}}{g^{(n-1)}} \right). \end{aligned}$$

Hence by iteration

$$(8.8) \qquad \frac{h_n - g^{(n)}}{h_n - p^{(n)}} = \prod_{\nu=1}^{n} \left(\frac{p^{\nu-1)}}{g^{(\nu-1)}} \right).$$

References.

1. Hayden, T. L. Continued fraction approximation to functions, *Numer. Math.* 7 (1965), 292–309.

2. Jacobsen, Lisa. General convergence of continued fractions, *Trans. Amer. Math. Soc.* 294 (1986), 477–485.

3. Jacobsen, Lisa and Thron, W. J. Limiting structures for sequences of linear fractional transformations, *Proc. Amer. Math. Soc.* 99 (1987), 141–146.

4. Jacobsen, Lisa and Waadeland, Haakon. Some useful formulas involving tails of continued fractions, *Lecture Notes in Mathematics No. 932*, Springer, Berlin 1982, pp. 99–105.

5. Jones, W. B. and Thron, W. J. *Continued Fractions, Analytic Theory and Applications*, Addison–Wesley, Reading, Mass. 1980.

6. Jones, W. B. and Thron, W. J. Continued fractions in numerical analysis, *Appl. Numer. Math.* 4 (1988), 143–230.

7. Lange, L. J. Lecture at Boulder Conference, June 1988.

8. Overholdt, Marius. A class of element and value regions for continued fractions, *Lecture Notes in Mathematics No. 932*, Springer, Berlin, 1982, pp. 194–205.

9. Thron, W. J. Another look at the definition of strong convergence of continued fractions, *Kongl. Norske Vidensk. Selskab. Skrifter* 1983–No. 1, pp. 128–137.

10. Thron, W. J. and Waadeland, Haakon. Accelerating convergence of limit periodic continued fractions, *Numer. Math.* 34 (1980), 155–170.

11. Thron, W. J. and Waadeland, Haakon. Truncation error bounds for limit periodic continued fractions, *Math. of Comp.* 40 (1983), 589–597.

12. Waadeland, Haakon. Some recent results in the analytic theory of continued fractions, *Nonlinear Numerical Methods and Rational Approximation*, Reidel, Amsterdam 1988, pp. 299–333.

13. Wall, H. S. On continued fractions and cross-ratio groups of Cremona transformations, *Bull. Amer. Math. Soc.* 40 (1934), 587–592.

14. Jacobsen, Lisa. Composition of linear fractional transformations in terms of tail sequences, *Proc. Amer. Math. Soc.* 97 (1986), 97–104.

15. Niethammer, W. and Wietschorke. On the acceleration of limit periodic continued fractions, *Numer. Math.* 44 (1984), 129–137.

16. Phipps, T. E. Jr. A continued fraction representation of eigenvalues, *SIAM Rev.* 13 (1971), 390–395.

Department of Mathematics
University of Colorado
Boulder, CO 80309-0426
U.S.A.

BOUNDARY VERSIONS OF WORPITZKY'S THEOREM AND OF PARABOLA THEOREMS

Haakon Waadeland
Department of Mathematics and Statistics
The University of Trondheim AVH
N–7055 Dragvoll
Norway

Abstract. What happens to the limit regions in Worpitzky's Theorem and in Parabola Theorems when the element regions are replaced by their boundaries? The present paper gives some answers to such questions.

The idea of "bestness" is of some interest in the analytic theory of continued fractions. The *best value region* V corresponding to a given element region E is defined in [2, p. 64] (in a more general setting). The *best limit region* L corresponding to E (see e. g. [5]) is essentially the set of all possible values of the continued fractions with all elements in E. The *best element region* E belonging to ([2, p. 64]) a given value region V or limit region L is the *largest* element set E belonging to the given value region or limit region. (Following [2, p. 64] the term *region* is here used loosely to mean any subset of **C**.)

On the "value side" bestness means essentially that the set is as *small* as possible, whereas on the "element side" it means that the set is as *large* as possible.

One could also think of another kind of bestness: To a given element region E for continued fractions

$$(1) \qquad \overset{\infty}{\underset{n=1}{\mathrm{K}}} \frac{a_n}{1},$$

let L be the best limit region. Then all points of L, except possibly 0, are values of some continued fraction (1) with all $a_n \in E$. The following question then arises quite naturally: Is it possible to reduce E to some $E' \subset E$, and still "fill the whole set" L? Is there a smallest such set E'? We could even go further, and be satisfied with a limit set L', smaller than L, but dense in L. A simple example of this is the following, described in [5]:

For $0 < p < q$,

$$X = \frac{p}{1} + \frac{q}{1} + \frac{p}{1} + \frac{q}{1} + \cdots = \frac{1}{2}\left(\sqrt{(1+p+q)^2 - 4pq} - 1 - q + p\right),$$

$$Y = \frac{q}{1} + \frac{p}{1} + \frac{q}{1} + \frac{p}{1} + \cdots = \frac{1}{2}\left(\sqrt{(1+p+q)^2 - 4pq} - 1 + q - p\right),$$

and

$$E = [p, q], L = [X, Y],$$

the interval L is the best limit region corresponding to the element region (interval) E. If E is replaced by the set

$$E' = \{p, q\} = \partial E,$$

consisting only of the end-points, then the best limit region L' is dense in L, provided that

$$pq \geq p + q.$$

In this case the boundary of E turns out to be of great significance for the limit region L. Also in more important cases the boundary of E provides significant information about L. For parabolic element regions P, described in e. g. [3], [4] and [6], the boundary of the best limit region merely consists of points representing values of certain 2-periodic continued fractions with elements from ∂P. Illustrations are given in [5].

In the present paper we shall look at two classical cases of closed element regions and best limit regions for continued fractions (1). We shall replace the original element region E by ∂E and determine the best corresponding limit region. It will turn out, that in one of the two cases to be studied the best limit region is a proper subset of the original one (and not dense in it), whereas in the other case it coincides with the original best limit region.

The first example is inspired by the Worpitzky Theorem [6], but in a slightly more general form than the classical one.

Theorem 1. *Let ρ be a fixed positive number, $0 < \rho \leq \frac{1}{2}$, and let F_ρ be the family of continued fractions*

$$(1) \qquad\qquad \overset{\infty}{\underset{n=1}{K}} \frac{a_n}{1},$$

defined by the condition

$$(2) \qquad\qquad |a_n| = \rho(1 - \rho)$$

for all n.

Then the set of all possible values f of continued fractions in F_ρ is the annulus A_ρ, given by

$$\rho\frac{1 - \rho}{1 + \rho} \leq |f| \leq \rho.$$

Proof: Let f_o be a possible value. Then all values f with $|f| = |f_o|$ are possible continued fraction values in F_ρ. Hence the set of values must be a disk or an annulus, in both cases centered at the origin. From Worpitzky's theorem follows that this disk or annulus must be contained in the disk

$$(4) \qquad\qquad |f| \leq \rho.$$

We shall first prove that the set of all values must be contained in A_ρ: Any continued fraction in F_ρ can be written in the form

$$\frac{\rho(1 - \rho)e^{i\theta}}{1 + g},$$

where $g \in F_\rho$. Since $|g| \leq \rho$ it follows that for any value f of a continued fraction in F_ρ we have

(5)
$$|f| \geq \frac{\rho(1-\rho)}{1+\rho}.$$

That this is sharp, follows from the fact that

$$\rho = \frac{\rho(1-\rho)}{1} + \frac{-\rho(1-\rho)}{1} + \frac{-\rho(1-\rho)}{1} + \frac{-\rho(1-\rho)}{1} + \cdots,$$

and that the right-hand side is in F_ρ.

We next prove that A_ρ is contained in the set of values of continued fractions in F_ρ: Since ρ is a value, all points on the circle

$$|w| = \rho$$

are values. By $\omega = 1/(1+w)$ this circle is mapped onto the circle

$$\left| \omega - \frac{1}{1-\rho^2} \right| = \frac{\rho}{1-\rho^2}.$$

(Fig. 1, left part.)

By $\omega \mapsto \rho(1-\rho)e^{i\theta}\omega$ for all $\theta \in [0, 2\pi)$ we get all points in the annulus

$$\frac{\rho(1-\rho)}{1+\rho} \leq |f| \leq \rho.$$

(Fig. 1, right part.)

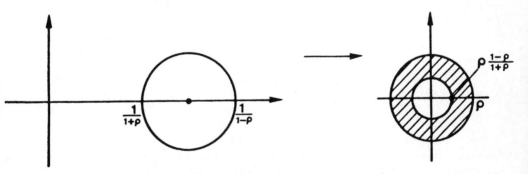

Fig. 1.

Hence A_ρ is contained in the set of continued fraction values for F_ρ, and the theorem is thus proved. Observe that in the classical case ($\rho = 1/2$) the annulus is $1/6 \leq |f| \leq 1/2$.

The next example is inspired by the one-parameter parabola theorem (actually only the value/limit region part of it).

Theorem 2. Let α be a fixed real number $-\pi/2 < \alpha < \pi/2$, and let P_α be the family of continued fractions

(1)
$$\overset{\infty}{\underset{n=1}{\mathrm{K}}} \frac{a_n}{1},$$

defined by the condition

(6)
$$|a_n| - \Re(a_n e^{-2i\alpha}) = \frac{1}{2} \cos^2 \alpha$$

for all n. Then the set of all possible values f of continued fractions in P_α is the halfplane V_α, given by

(7)
$$\Re(f e^{-i\alpha}) \geq -\frac{1}{2} \cos \alpha,$$

minus the origin.

Proof: It is well known that the set of all values of all continued fractions in P_α contains the straight line L_α given by
$$\Re(f e^{-i\alpha}) = -\frac{1}{2} \cos \alpha.$$

In fact, a certain subfamily of the family of 2-periodic continued fractions in P_α (and indeed av very slim one) suffices, see [3], [4] and [5].

By the transformation
$$\omega = \frac{1}{1 + w}$$

the line L_α is mapped onto the circle with center at $e^{-i\alpha}/\cos \alpha$ and radius $1/\cos \alpha$ (minus the origin). (Fig. 2 and 3.)

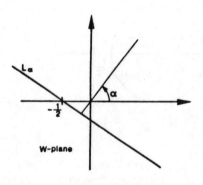

Fig. 2.

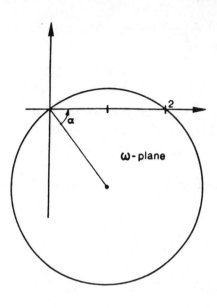

Fig. 3.

This circle (fig. 3) can be described by the following representation

(8)
$$\omega = \frac{2\cos\theta}{\cos\alpha}e^{i(\theta-\alpha)}, -\pi/2 < \theta < \pi/2.$$

An arbitrary point on the parabola is

(9)
$$a_n = \frac{\cos^2\alpha}{2(1-\cos\phi)}e^{i(\phi+2\alpha)}, 0 < \phi < 2\pi.$$

The set of all possible continued fraction values in P_α must contain all values

$$\frac{a_n}{1+w} = a_n\omega,$$

where ω is as in (8) and a_n as in (9), i. e. all values

(10)
$$e^{i\alpha}\cos\alpha\frac{\cos\theta}{(1-\cos\phi)}e^{i(\phi+\theta)},$$

where $-\pi/2 < \theta < \pi/2, 0 < \phi < 2\pi$.

Here we shall use the following lemma:

Lemma 3. Let $-\pi/2 < \theta < \pi/2$ and $0 < \phi < 2\pi$ be independent variables in their intervals, and let G be given by

$$(11) \qquad G(\phi, \theta) = \frac{2\cos\theta}{1 - \cos\phi} e^{i(\theta + \phi)}.$$

Then the range of G is exactly the half plane $\Re(G) \geq -1$ minus the origin.

Proof. Clearly the range of G is symmetric with respect to the real axis, since

$$G(2\pi - \phi, -\theta) = \overline{G(\phi, \theta)}.$$

Furthermore, the "natural extension" of $G(\phi, \theta)$ to all ϕ-values $\neq 2k\pi$, $k \in \mathbf{Z}$, by

$$G(\phi, \theta) = G(2\pi + \phi, \theta),$$

does not change the range.

For a fixed $\eta = \theta + \phi$, $0 \leq \eta < \pi/2$ we have

$$G(\phi, \theta) = G(\eta - \theta, \theta) = \frac{2\cos\theta}{1 - \cos(\eta - \theta)} e^{i\eta}.$$

Let θ increase from η to $\pi/2$ (and simultaneously $\phi = \eta - \theta$ decrease from 0 to $-(\pi/2 - \eta)$). Then

$$\frac{2\cos\theta}{1 - \cos(\eta - \theta)}$$

decreases from ∞ to 0. Hence the whole ray $\arg\omega = \eta$ is contained in the range. Since η is arbitrary in $[0, \pi/2)$ and the range is symmetric with respect to the real axis we have established that the half plane $\Re\omega > 0$ is contained in the range.

For $\theta + \phi = \pi/2$ we have

$$G(\phi, \theta) = G(\phi, \pi/2 - \phi) = \frac{2\sin\phi}{1 - \cos\phi} i = 2i\cot\phi/2.$$

Let ϕ increase from 0 to π (and $\theta = \pi/2 - \phi$ decrease from $\pi/2$ to $-\pi/2$). Then $2\cot\phi/2$ decreases from ∞ to 0. Hence the positive imaginary axis and, by symmetry, the whole imaginary axis minus the origin is contained in the range of G.

Finally fix an arbitrary $\eta = \theta + \phi$ in the interval $\pi/2 < \eta \leq \pi$. Then

$$(12) \qquad G(\phi, \theta) = G(\phi, \eta - \phi) = \frac{2\cos(\eta - \phi)}{1 - \cos\phi} e^{i\eta}.$$

Here the permitted ϕ-values between 0 and 2π are

$$(13) \qquad \eta - \frac{\pi}{2} < \phi < \eta + \frac{\pi}{2},$$

since $\theta = \eta - \phi$ is required to be between $-\pi/2$ and $\pi/2$. The function is continuos and positive, and tends to 0 when ϕ tends to $\eta - \pi/2$ or $\eta + \pi/2$. Hence it has a positive maximum $M(\eta)$, and the range of the function

$$\frac{2\cos(\eta - \phi)}{1 - \cos\phi}$$

is the interval $(0, M(\eta)]$. Now

$$\Re\left(\frac{2\cos(\eta - \phi)}{1 - \cos\phi}e^{i\eta}\right) = \frac{2\cos(\eta - \phi)\cos\eta}{1 - \cos\phi} =$$

$$= \frac{\cos(2\eta - \phi) + \cos\phi}{1 - \cos\phi} = -1 + \frac{1 + \cos(2\eta - \phi)}{1 - \cos\phi} \geq -1,$$

where $=$ holds iff $2\eta - \phi = \pi$, which is permitted, since we have (from (13))

$$\eta - \frac{\pi}{2} < 2\eta - \phi < \eta + \frac{\pi}{2}.$$

This shows that $M(\eta)\cos\eta = -1$. Then also the strip $-1 \leq \Re w < 0$ is in the range of G. This concludes the proof of the lemma.

From Lemma 3 the rest of the proof of Theorem 2 follows almost immediately, since (10) may be written

(10')
$$\frac{e^{i\alpha}\cos\alpha}{2}G(\phi, \theta),$$

and the range of this is exactly the half plane V_α, given by (7), minus the origin. Actually, we have so far only proved that the set of all possible values f of continued fractions in P_α *contains* V_α minus the origin, but since (from the Parabola Theorem) it must be *contained* *in* V_α minus the origin, the proof of Theorem 2 is established.

Remark. Lemma 3 is a strengthened form of Lemma 2.1 in [3], which in turn is a strengthened form of a lemma in [6]. Our proof of Lemma 3 is closely related to the proof of Lemma 2.1, but in the latter case the starshape of the range followed trivially from the presence of a factor t, varying in $[0, 1]$ independently of the two other parameters.

Final remarks. Although we have been able to reduce the element region from a parabolic region down to its boundary (6) without reducing the limit region (7), we have not established "bestness" in the sense presented in the introduction, i. e. we have not proved, that furter reduction of the element region will reduce the set of continued fraction values.

In the proof of more general parabola theorems (see e. g. [3]) Lemma 2.1 in [3] may be replaced by Lemma 3, which opens up for the possibility of boundary versions. Another interesting question is to ask what happens when the oval regions in the oval theorems [1] are replaced by their boundaries. In view of the result in the Worpitzky case and the role of the Worpitzky element disk and the parabolic region as extreme cases of oval regions it is likely that the limit region will have a hole, which degenerates to a point (origin) in the case of a parabolic region. Even for more general element/ limit regions it seems to be worthwhile to look into what happens if the (closed) element region is reduced to its boundary. At least some tempting guesses are lying in that area.

References

1. L. Jacobsen and W. J. Thron, *Oval convergence regions and circular limit regions for continued fractions* $K(a_n/1)$, Analytic Theory of Continued Fractions II (W. J. Thron, Ed.), Lecture Notes in Mathematics 1199, Springer-Verlag, Berlin (1986), 90–126.

2. W. B. Jones and W. J. Thron, "Continued fractions: Analytic theory and applications, Encyclopedia of Mathematics and its Applications, 11," Addison-Wesley, now available from Cambridge University Press, 1980.

3. W. B. Jones, W. J. Thron and H. Waadeland, *Value regions for continued fractions* $K(a_n/1)$ *whose elements lie in parabolic regions*, Math. Scand. 56 (1985), 5–14.

4. W. Leighton and W. J. Thron, *On value regions of continued fractions*, Bull. Amer. Math. Soc. 48, No. 12 (1942), 917–920.

5. E. Rye and H. Waadeland, *Reflections on value regions, limit regions and truncation errors for continued fractions*, Numer. Math. 47 (1985), 191–215.

6. W. J. Thron, *On parabolic convergence regions for continued fractions*, Math. Z. 69 (1958), 173–182.

7. J. D. T. Worpitzky, *Untersuchungen über die Entwickelung der monodromen und monogenen Funktionen durch Kettenbrüche*, Friedrichs- Gymnasium und Realschule Jahresbericht, Berlin (1865), 3–39.

LECTURE NOTES IN MATHEMATICS
Edited by A. Dold and B. Eckmann

Some general remarks on the publication of proceedings of congresses and symposia

Lecture Notes aim to report new developments - quickly, informally and at a high level. The following describes criteria and procedures which apply to proceedings volumes. The editors of a volume are strongly advised to inform contributors about these points at an early stage.

§1. One (or more) expert participant(s) of the meeting should act as the responsible editor(s) of the proceedings. They select the papers which are suitable (cf. §§ 2, 3) for inclusion in the proceedings, and have them individually refereed (as for a journal). It should not be assumed that the published proceedings must reflect conference events faithfully and in their entirety. Contributions to the meeting which are not included in the proceedings can be listed by title. The series editors will normally not interfere with the editing of a particular proceedings volume - except in fairly obvious cases, or on technical matters, such as described in §§ 2, 3. The names of the responsible editors appear on the title page of the volume.

§2. The proceedings should be reasonably homogeneous (concerned with a limited area). For instance, the proceedings of a congress on "Analysis" or "Mathematics in Wonderland" would normally not be sufficiently homogeneous.

One or two longer survey articles on recent developments in the field are often very useful additions to such proceedings - even if they do not correspond to actual lectures at the congress. An extensive introduction on the subject of the congress would be desirable.

§3. The contributions should be of a high mathematical standard and of current interest. Research articles should present new material and not duplicate other papers already published or due to be published. They should contain sufficient information and motivation and they should present proofs, or at least outlines of such, in sufficient detail to enable an expert to complete them. Thus resumes and mere announcements of papers appearing elsewhere cannot be included, although more detailed versions of a contribution may well be published in other places later.

Surveys, if included, should cover a sufficiently broad topic, and should in general not simply review the author's own recent research. In the case of surveys, exceptionally, proofs of results may not be necessary.

"Mathematical Reviews" and "Zentralblatt für Mathematik" require that papers in proceedings volumes carry an explicit statement that they are in final form and that no similar paper has been or is being submitted elsewhere, if these papers are to be considered for a review. Normally, papers that satisfy the criteria of the Lecture Notes in Mathematics series also satisfy this

.../...

requirement, but we would strongly recommend that the contributing authors be asked to give this guarantee explicitly at the beginning or end of their paper. There will occasionally be cases where this does not apply but where, for special reasons, the paper is still acceptable for LNM.

§4. Proceedings should appear soon after the meeeting. The publisher should, therefore, receive the complete manuscript within nine months of the date of the meeting at the latest.

§5. Plans or proposals for proceedings volumes should be sent to one of the editors of the series or to Springer-Verlag Heidelberg. They should give sufficient information on the conference or symposium, and on the proposed proceedings. In particular, they should contain a list of the expected contributions with their prospective length. Abstracts or early versions (drafts) of some of the contributions are very helpful.

§6. Lecture Notes are printed by photo-offset from camera-ready typed copy provided by the editors. For this purpose Springer-Verlag provides editors with technical instructions for the preparation of manuscripts and these should be distributed to all contributing authors. Springer-Verlag can also, on request, supply stationery on which the prescribed typing area is outlined. Some homogeneity in the presentation of the contributions is desirable.

Careful preparation of manuscripts will help keep production time short and ensure a satisfactory appearance of the finished book. The actual production of a Lecture Notes volume normally takes 6 –8 weeks.

Manuscripts should be at least 100 pages long. The final version should include a table of contents and as far as applicable a subject index.

§7. Editors receive a total of 50 free copies of their volume for distribution to the contributing authors, but no royalties. (Unfortunately, no reprints of individual contributions can be supplied.) They are entitled to purchase further copies of their book for their personal use at a discount of 33.3 %, other Springer mathematics books at a discount of 20 % directly from Springer-Verlag. Contributing authors may purchase the volume in which their article appears at a discount of 33.3 %.

Commitment to publish is made by letter of intent rather than by signing a formal contract. Springer-Verlag secures the copyright for each volume.

Vol. 1290: G. Wüstholz (Ed.), Diophantine Approximation and Transcendence Theory. Seminar, 1985. V, 243 pages. 1987.

Vol. 1291: C. Mœglin, M.-F. Vignéras, J.-L. Waldspurger, Correspondances de Howe sur un Corps p-adique. VII, 163 pages. 1987

Vol. 1292: J.T. Baldwin (Ed.), Classification Theory. Proceedings, 1985. VI, 500 pages. 1987.

Vol. 1293: W. Ebeling, The Monodromy Groups of Isolated Singularities of Complete Intersections. XIV, 153 pages. 1987.

Vol. 1294: M. Quøffélec, Substitution Dynamical Systems – Spectral Analysis. XIII, 240 pages. 1987.

Vol. 1295: P. Lelong, P. Dolbeault, H. Skoda (Réd.), Séminaire d'Analyse P. Lelong – P. Dolbeault – H. Skoda. Seminar, 1985/1986. VII, 283 pages. 1987.

Vol. 1296: M.-P. Malliavin (Ed.), Séminaire d'Algèbre Paul Dubreil et Marie-Paule Malliavin. Proceedings, 1986. IV, 324 pages. 1987.

Vol. 1297: Zhu Y.-l., Guo B.-y. (Eds.), Numerical Methods for Partial Differential Equations. Proceedings. XI, 244 pages. 1987.

Vol. 1298: J. Aguadé, R. Kane (Eds.), Algebraic Topology, Barcelona 1986. Proceedings. X, 255 pages. 1987.

Vol. 1299: S. Watanabe, Yu.V. Prokhorov (Eds.), Probability Theory and Mathematical Statistics. Proceedings, 1986. VIII, 589 pages. 1988.

Vol. 1300: G.B. Seligman, Constructions of Lie Algebras and their Modules. VI, 190 pages. 1988.

Vol. 1301: N. Schappacher, Periods of Hecke Characters. XV, 160 pages. 1988.

Vol. 1302: M. Cwikel, J. Peetre, Y. Sagher, H. Wallin (Eds.), Function Spaces and Applications. Proceedings, 1986. VI, 445 pages. 1988.

Vol. 1303: L. Accardi, W. von Waldenfels (Eds.), Quantum Probability and Applications III. Proceedings, 1987. VI, 373 pages. 1988.

Vol. 1304: F.Q. Gouvêa, Arithmetic of p-adic Modular Forms. VIII, 121 pages. 1988.

Vol. 1305: D.S. Lubinsky, E.B. Saff, Strong Asymptotics for Extremal Polynomials Associated with Weights on ℝ. VII, 153 pages. 1988.

Vol. 1306: S.S. Chern (Ed.), Partial Differential Equations. Proceedings, 1986. VI, 294 pages. 1988.

Vol. 1307: T. Murai, A Real Variable Method for the Cauchy Transform, and Analytic Capacity. VIII, 133 pages. 1988.

Vol. 1308: P. Imkeller, Two-Parameter Martingales and Their Quadratic Variation. IV, 177 pages. 1988.

Vol. 1309: B. Fiedler, Global Bifurcation of Periodic Solutions with Symmetry. VIII, 144 pages. 1988.

Vol. 1310: O.A. Laudal, G. Pfister, Local Moduli and Singularities. V, 117 pages. 1988.

Vol. 1311: A. Holme, R. Speiser (Eds.), Algebraic Geometry, Sundance 1986. Proceedings. VI, 320 pages. 1988.

Vol. 1312: N.A. Shirokov, Analytic Functions Smooth up to the Boundary. III, 213 pages. 1988.

Vol. 1313: F. Colonius, Optimal Periodic Control. VI, 177 pages. 1988.

Vol. 1314: A. Futaki, Kähler-Einstein Metrics and Integral Invariants. IV, 140 pages. 1988.

Vol. 1315: R.A. McCoy, I. Ntantu, Topological Properties of Spaces of Continuous Functions. IV, 124 pages. 1988.

Vol. 1316: H. Korezlioglu, A.S. Ustunel (Eds.), Stochastic Analysis and Related Topics. Proceedings, 1986. V, 371 pages. 1988.

Vol. 1317: J. Lindenstrauss, V.D. Milman (Eds.), Geometric Aspects of Functional Analysis. Seminar, 1986–87. VII, 289 pages. 1988.

Vol. 1318: Y. Felix (Ed.), Algebraic Topology – Rational Homotopy. Proceedings, 1986. VIII, 245 pages. 1988

Vol. 1319: M. Vuorinen, Conformal Geometry and Quasiregular Mappings. XIX, 209 pages. 1988.

Vol. 1320: H. Jürgensen, G. Lallement, H.J. Weinert (Eds.), S： groups, Theory and Applications. Proceedings, 1986. X, 416 pa； 1988.

Vol. 1321: J. Azéma, P.A. Meyer, M. Yor (Eds.), Séminaire Probabilités XXII. Proceedings. IV, 600 pages. 1988.

Vol. 1322: M. Métivier, S. Watanabe (Eds.), Stochastic Analy Proceedings, 1987. VII, 197 pages. 1988.

Vol. 1323: D.R. Anderson, H.J. Munkholm, Boundedly Contr Topology. XII, 309 pages. 1988.

Vol. 1324: F. Cardoso, D.G. de Figueiredo, R. Iório, O. Lopes (E Partial Differential Equations. Proceedings, 1986. VIII, 433 pa 1988.

Vol. 1325: A. Truman, I.M. Davies (Eds.), Stochastic Mechanics Stochastic Processes. Proceedings, 1986. V, 220 pages. 1988.

Vol. 1326: P.S. Landweber (Ed.), Elliptic Curves and Modular Form Algebraic Topology. Proceedings, 1986. V, 224 pages. 1988.

Vol. 1327: W. Bruns, U. Vetter, Determinantal Rings. VII,236 pa 1988.

Vol. 1328: J.L. Bueso, P. Jara, B. Torrecillas (Eds.), Ring The Proceedings, 1986. IX, 331 pages. 1988.

Vol. 1329: M. Alfaro, J.S. Dehesa, F.J. Marcellan, J.L. Rubio Francia, J. Vinuesa (Eds.): Orthogonal Polynomials and their App tions. Proceedings, 1986. XV, 334 pages. 1988.

Vol. 1330: A. Ambrosetti, F. Gori, R. Lucchetti (Eds.), Mathema Economics. Montecatini Terme 1986. Seminar. VII, 137 pages. 19

Vol. 1331: R. Bamón, R. Labarca, J. Palis Jr. (Eds.), Dynam Systems, Valparaiso 1986. Proceedings. VI, 250 pages. 1988.

Vol. 1332: E. Odell, H. Rosenthal (Eds.), Functional Analysis. F ceedings, 1986–87. V, 202 pages. 1988.

Vol. 1333: A.S. Kechris, D.A. Martin, J.R. Steel (Eds.), Cabal Sem 81–85. Proceedings, 1981–85. V, 224 pages. 1988.

Vol. 1334: Yu.G. Borisovich, Yu. E. Gliklikh (Eds.), Global Analy – Studies and Applications III. V, 331 pages. 1988.

Vol. 1335: F. Guillén, V. Navarro Aznar, P. Pascual-Gainza, F. Pue Hyperrésolutions cubiques et descente cohomologique. XII, 1 pages. 1988.

Vol. 1336: B. Helffer, Semi-Classical Analysis for the Schrödin Operator and Applications. V, 107 pages. 1988.

Vol. 1337: E. Sernesi (Ed.), Theory of Moduli. Seminar, 1985. VIII, 2 pages. 1988.

Vol. 1338: A.B. Mingarelli, S.G. Halvorsen, Non-Oscillation Doma of Differential Equations with Two Parameters. XI, 109 pages. 1988.

Vol. 1339: T. Sunada (Ed.), Geometry and Analysis of Manifold Procedings, 1987. IX, 277 pages. 1988.

Vol. 1340: S. Hildebrandt, D.S. Kinderlehrer, M. Miranda (Eds Calculus of Variations and Partial Differential Equations. Proceeding 1986. IX, 301 pages. 1988.

Vol. 1341: M. Dauge, Elliptic Boundary Value Problems on Corr Domains. VIII, 259 pages. 1988.

Vol. 1342: J.C. Alexander (Ed.), Dynamical Systems. Proceeding 1986–87. VIII, 726 pages. 1988.

Vol. 1343: H. Ulrich, Fixed Point Theory of Parametrized Equivari Maps. VII, 147 pages. 1988.

Vol. 1344: J. Král, J. Lukeš, J. Netuka, J. Veselý (Eds.), Poten Theory – Surveys and Problems. Proceedings, 1987. VIII, 271 pag 1988.

Vol. 1345: X. Gomez-Mont, J. Seade, A. Verjovski (Eds.), Holomorp Dynamics. Proceedings, 1986. VII, 321 pages. 1988.

Vol. 1346: O. Ya. Viro (Ed.), Topology and Geometry – Roh Seminar. XI, 581 pages. 1988.

Vol. 1347: C. Preston, Iterates of Piecewise Monotone Mappings an Interval. V, 166 pages. 1988.

Vol. 1348: F. Borceux (Ed.), Categorical Algebra and its Applicatio Proceedings, 1987. VIII, 375 pages. 1988.

Vol. 1349: E. Novak, Deterministic and Stochastic Error Bounds Numerical Analysis. V, 113 pages. 1988.